Software Engineering Practice

Software Engineering Practice
A Case Study Approach

Thomas B. Hilburn
Massood Towhidnejad

CRC Press
Taylor & Francis Group
Boca Raton London New York

CRC Press is an imprint of the
Taylor & Francis Group, an **informa** business

A CHAPMAN & HALL BOOK

First Edition published 2021
by CRC Press
6000 Broken Sound Parkway NW, Suite 300, Boca Raton, FL 33487-2742

and by CRC Press
2 Park Square, Milton Park, Abingdon, Oxon, OX14 4RN

© 2021 Taylor & Francis Group, LLC

CRC Press is an imprint of Taylor & Francis Group, LLC

ISBN: 978-1-4665-9167-7 (hbk)
ISBN: 978-0-429-16849-9 (ebk)

Typeset in Garamond
by codeMantra

Contents

Preface ix
Acknowledgments xiii
Authors xv

1 In The Beginning...**1**
　　Birth of *DigitalHome* 1
　　Forming a Project Team...10
　　Assessing *DigitalHome's* Needs...12
　　　　DH Customer Need Statement 13
　　　　DH High-Level Requirements Definition 13
　　Case Study Exercises...17

2 Launching *DigitalHome* **23**
　　Project Launch...23
　　Team Building...25
　　Software Development Process 29
　　Development Strategy.. 42
　　Case Study Exercises...45

3 Assuring *DigitalHome* Quality **49**
　　Software Quality Assurance..49
　　Software Quality Assurance Processes 58
　　Quality Measurement and Defect Tracking....................................60
　　Case Study Exercises...79

4 Managing the DH Project **91**
　　Project Planning 91
　　Planning Activities..96
　　Risk Management...105
　　Software Configuration Management.. 110
　　Quality Planning... 116
　　Case Study Exercises... 117

5   Engineering the DH Requirements 123
   Software Requirements Fundamentals...123
   Eliciting Requirements 129
   Analyzing Requirements...132
   Specifying Requirements 137
   Validating Requirements... 141
   Case Study Exercises...143

6   Designing *DigitalHome* 151
   Software Design Concepts and Principles... 151
   Software Architecture 159
   Architecture Views and Styles.. 161
   Object-Oriented Design 171
   Design Verification 181
   Software Reuse and Design Patterns...184
   Documenting Software Design..189
   Case Study Exercises.. 191

7   Constructing *DigitalHome* 201
   Build/Integration Plan...201
   Construction Fundamentals 206
   Unit Construction 208
   Case Study Exercises..216

8   Maintaining *DigitalHome* 221
   Maintenance Fundamentals..221
   Maintenance Processes.. 226
   Maintenance Techniques 228
   Case Study Exercises..231

9   Acting Ethically and Professionally 237
   Software Engineering Professional Issues...237
   Code of Ethics and Professional Conduct..243
   Software Development Standards 245
   Software Legal Issues...247
   Case Study Exercises..250

10   Using the Scrum Development Process 255
   Scrum Process Overview..255
   Backlog Generation and Grooming 260
   Building the Product 267
   Scrum Reflection Activities..272
   Case Study Exercises..275

References 279

Appendix A: Digital Home Customer Need Statement 283

Appendix B: *DigitalHome* Software Requirements Specification...............287

Appendix C: *DigitalHome* Use Case Model 301

Index 331

Preface

Goals of This Book

This book is envisioned as a new approach to learning about software engineering theory and practice. Specifically, the book is designed to:

- Support teaching software engineering using a comprehensive case study covering the complete software development life cycle
- Offer opportunities for students to actively learn about and engage in software engineering practice
- Provide a realistic environment to study a wide array of software engineering topics through a full software development life cycle

Why Another Book on Software Engineering?

Software is an important part of almost every aspect of human endeavor today. Software engineering methods and technology have advanced greatly in the last 30 years. Professionals with software engineering knowledge and skills are in high demand.

There are many fine books that address the importance of software engineering and how to prepare students for a professional career in software engineering. Some provide an overview of the range of software engineering topics covering the complete software development life cycle. Others provide depth about specific areas of software engineering practice (e.g., software requirements, software design, software construction, software testing, and software process).

This book provides a broad discussion of the topics covering the entire software development life cycle using a comprehensive case study that addresses each topic discussed. The book includes the following:

- a description of the development, by the fictional company Homeowner, of a *DigitalHome* (DH) System, a system with "smart" devices for controlling home lighting, temperature, humidity, small appliance power, and security;

- a set of scenarios that provide a realistic framework for use of the DH System material;
- just-in-time training: each chapter includes mini tutorials introducing various software engineering topics that are discussed in that chapter and used in the case study; and
- a set of case study exercises that provide an opportunity to engage students in software development practice, either individually or in a team environment.

Projected Audience and Use

This book's primary intended use is as the course textbook in an Introductory Overview Software Engineering course. It could be supplemented with reading from the references provided throughout the book.

It is assumed that the audience has some basic understanding of the nature of computing and can use fundamental program constructs (sequence, selection, and repetition) and write simple programs in a high-level object-oriented programming language (such as Java).

The book might also be used to supplement instruction in a variety of other computing courses, both at the undergraduate and graduate level:

- Programming Courses
- Object-Oriented Analysis and Design
- Software Requirements
- Software Quality or Software Verification & Validation
- Software Project Management
- Computing Capstone Project Courses (and other project courses)
- Software Maintenance
- Database Design
- Simulation and Modeling
- Software Engineering Professional Practice

As described above, faculty can use the book throughout a computing curriculum. The book has great flexibility in structure and content. After the first chapter, material in each chapter is relatively self-contained and can be used as needed and appropriate. For example, a course on Software Requirements Engineering might just use Chapters 1, 2, 3, 5, and 10.

The DH System material can be adapted and enhanced to suit individual program or course needs. Although the book discusses a set of artifacts that cover a full development life cycle, the project itself (*DigitalHome*) is designed so that it can be expanded in scope with almost no limitation, thereby allowing the faculty adopting the book to use the material for many semesters and not worry about exercise repetitions.

Also, the book can be used to support professional development and/or certification of practicing software engineers. The case study exercises could be integrated with presentations in a workshop or short course for professionals.

Education and Pedagogy Contributions

The book supports a student-centered, "active" learning style of teaching.

- The DH case study exercises provide a variety of opportunities for students to engage in realistic activities related to the theory and practice of software engineering.
- Although the case method of teaching (the use of case studies as the principal teaching paradigm) has proven its worth and is a widely used method of teaching in fields such as business, law, and medicine [Herreid 1994], it is yet to be accepted to the same extent in engineering education, except in the area of engineering ethics.
- The text uses a fictitious team of software engineers to portray the nature of software engineering and to depict what actual engineers do when practicing software engineering. The team members have defined roles and responsibilities, diverse and realistic backgrounds and personalities, and are intended to engage and motivate students with real-world experiences. The project team is made up of a varied set of developers: Disha Chandra, Michel Jackson, Georgia Magee, and Massood Zewail, with Jose Ortiz as their Division head.
- All of the DH cases study exercises can be used as team/group exercises in collaborative learning. Many of the exercises have specific goals related to team building and teaming skills.
- Each case study exercise includes formal communication activities (oral and written) about the results of the exercise.
- The DH team creates a development process (Table 2.1), which includes elements from both the Unified incremental process [Scott 2002] and the V-Model [Pressman 2015]. There is no intent to advocate for this process over other models, such as spiral process. The intent is to provide a basis for discussion of all the key software engineering practices needed for the development of a large, complex software system.
- Given the popularity of the agile processes, the book also introduces the agile development life cycle and provides a detailed coverage of Scrum process. Again, there is no intent to advocate Scrum over other agile processes, and faculty teaching this class can easily replace Scrum by their favorite agile process.

Acknowledgments

This book is not only a product of the authors' efforts, but its contents and style have been influenced by the contributions of hundreds, maybe thousands, of others.

The list of references at the end the book is a good starting point. The authors of these works have provided us with the knowledge and support we needed to learn about, teach, and write about software engineering.

The students in the courses the authors have taught in computer science, computer engineering, and software engineering deserve special thanks. They have provided us with a testbed for almost all the ideas and approaches presented in this book. The authors have used the case method approach, along with the *DigitalHome* scenarios, exercises, and artifacts, in a variety of classes: introductory software engineering, software construction, object-oriented design, software architecture, software requirements, and software quality. Student performance and feedback has been invaluable in determining how best to explain and use the case method approach.

Initial work on the *DigitalHome* case study was funded as part of the NSF project: "The Network Community for Software Engineering Education" (SWENET) (NSF 0080502). Later work on the case study project was funded through NSF's (DUE-0941768) "Curriculum-wide Software Development Case Study". Also, under DUE-0941768, the authors delivered two faculty workshops on case study teaching, in which we used DH scenarios, exercises, and artifacts. The faculty attendees provided both real-time feedback during discussion sessions and a formal evaluation of the material at the end of the workshop. This gave us invaluable information about how to modify and assemble our material to best assist others teaching software development courses.

Finally, we offer our special thanks and appreciation to our esteemed colleague, Dr. Salamah Salamah, who influenced much of the style and content of this book. He conducted multiple reviews and provided valuable feedback throughout the development of the manuscript. We also extend our thanks to two of our former graduate students, Bapu Hirave and Langpap Bernd, who were so helpful in the development of the *DigitalHome* artifacts.

Authors

Dr. Thomas B. Hilburn is a Professor Emeritus of Software Engineering and a distinguished Engineering Professor at Embry-Riddle Aeronautical University, Daytona Beach, Florida, and was a Visiting Scientist at the Software Engineering Institute, Carnegie-Mellon University, Pittsburgh, Pennsylvania, from 1997 to 2009. He has worked on software engineering development, research, and education projects with the FAA, General Electric, Lockheed-Martin, the Harris Corp., The MITRE Corporation, DOD, FIPSE, the SEI, the NSF, the ACM, and the IEEE Computer Society. His interests include software processes, object-oriented analysis and design, formal specification techniques, and curriculum development, and he has published over 80 papers in these areas. He is an IEEE Certified Software Developer, an IEEE Software Engineering Certified Instructor, and has chaired committees on the Professional Activities Board and the Educational Activities Board of the IEEE Computer Society.

Dr. Massood Towhidnejad is Professor of Software Engineering at Embry-Riddle Aeronautical University, Daytona Beach, Florida. His research interests include Software Engineering, Software Quality Assurance and Testing, Autonomous Systems, Air Traffic Management, and STEM Education. He has worked on software engineering development and research projects with the NSF (National Science Foundation), NASA Goddard Space Flight Research Center, FAA (Federal Aviation Administration), NOAA (National Oceanic and Atmospheric Administration), and a number of corporations. He was a contributing author for Graduate Software Engineering Reference Curriculum (GSwE2009), Software Engineering Competency Model (SWECOM 2014), Graduate Reference Curriculum for Systems Engineering (GRCSE), and IEEE Certified Software Development Associate (CSDA) training materials. His work has been published in over 100 papers.

In addition to his university position, he served as a Visiting Research Associate at the FAA, Faculty Fellow at NASA Goddard Flight Research Center, and Software Quality Assurance Manager at Carrier Global Corporation.

Chapter 1

In The Beginning

Birth of *DigitalHome*

Although home building has its ups and downs, the long-term prospects call for increases in total home square footage and intensified interest by home owners in high technology devices. New home owners, especially young professionals, are interested in living in houses that have "smart" features that automate and facilitate energy efficiency, entertainment delivery, communication, household chores, and home security. The design and construction of such dwellings relies on the latest technology (devices with smart capabilities, distributed computing, wireless and web communications, intelligent agents, etc.).

In August 202X, *HomeOwner* held its annual strategic planning retreat. During these one-week sessions *HomeOwner* officers and other upper-level managers concentrate on new initiatives that will advance the *HomeOwner* position as the industry leader in household product retailing.

On the first day of the 202X retreat, Dick Punch (VP of Marketing) and Judy Fielder (Chief Information Officer) made a presentation about the trend for "smart" technology and the market for building dwellings that integrate smart technology into every aspect of home living (environmental and energy control, security, communication, entertainment, household chores, etc.). Punch and Judy argued that the potential market for "smart" houses is tremendous, and although *HomeOwner* currently sells a number of smart home devices, it does not market the "complete package". There are construction companies that during the planning and building phases incorporate computer and communication systems needed to combine and control the smart devices in such a way that best serves the needs and desires of today's new tech-savvy upwardly mobile professional. Punch and Judy recommended that *HomeOwner* become a competitor in this field, and look for opportunities to push the industry to building "smarter" houses.

Red Sharpson, CEO of HomeOwner, loved the smart house idea and decided to devote the rest of the week to formulating plans for this new initiative. After days of discussion and debate the following decisions were made:

- Establish a new *DigitalHomeOwner* Division of *HomeOwner* to be headed by Jose Ortiz (currently Deputy CIO).
- Develop a two-year strategic plan for the division.
- Assemble a staff for the division.
- Develop a one-year division budget plan.

In the next year, *DigitalHomeOwner* will carry out the following activities:

- With the assistance of the Marketing Division, *DigitalHomeOwner* will research the existing smart houses, and conduct a needs assessment for a *DigitalHome* product, which will provide the computer and communication infrastructure for managing and controlling the "smart" devices in a home to best meet the needs and desires of homeowners. In other words, building houses that take the advantage of Internet of Things (IoT) technology.

■ Choose a set of features for the *DigitalHome* product to develop a high-quality "proof of concept" prototype to illustrate how the *DigitalHome* needs will be met. Because of time constraints, the features chosen will be a subset of the actual needs. It is envisioned that this will primarily be a "software effort", with simulation being used to represent dwelling structures, hardware devices, and communication links.

■ Jose Ortiz will form a team and supervise it in the development of the *DigitalHome* prototype.

■ Jose Ortiz will deliver a report on the *DigitalHome* efforts at the August 202Y strategic planning retreat.

■ In addition, *DigitalHome* will include the following special considerations:

– It is anticipated that the *DigitalHome* prototype will be the foundation for future development of *DigitalHomeOwner* products. Hence, it is essential that the development team use established software engineering practices and fully document their work.

– In development of the prototype, *DigitalHomeOwner* should have two primary client groups: (1) potential *DigitalHome* users and (2) the upper management of *HomeOwner*. These clients should be part of an acceptance testing process for the prototype.

– At the retreat, Judy Fielder, the CIO, also presented a proposal (*HomeOwner Cloud Computing Initiative Proposal*) for development of a *HomeOwner* cloud server. Although the decision on the cloud initiative was tabled for further study, it was recommended that the *DigitalHome* developers consult the guidance provided in the proposal, especially as related to the privacy and security of client communication and data.

Judy Fielder pointed out that to make this project a success it was important that the project team use the best software engineering practices. Red Sharpson stated that he had little understanding about software engineering and did not know what the "best" software engineering practices were. Judy suggested that they ask Jose Ortiz to deliver a talk on software engineering on the last day of the retreat. Red agreed.

MINITUTORIAL 1.1: WHAT IS SOFTWARE ENGINEERING?

The *IEEE Standard Systems and Software Engineering Vocabulary* [IEEE 24765] defines *software engineering* as "the application of a systematic, disciplined, quantifiable approach to the development, operation, and maintenance of software; that is, the application of engineering to software". One of the earliest recorded uses of the term "Software Engineering" was in the 1968 NATO Software Engineering Conference; in the Conference report [Naur 1969], it was stated that the term was used to imply "the need for software manufacture to be based on the types of theoretical foundations and practical disciplines, that are traditional in the established branches of engineering". Hence, software engineering is a relatively young engineering discipline.

SOFTWARE ENGINEERING PROBLEMS

Software plays a critical and central role in almost all aspects of human endeavor. The size, complexity, and number of application domains of software products and services continue to grow. Unfortunately, there are serious problems in the cost, timeliness, and quality of development of many software systems. A significant number of development projects are never completed, and many of those completed do not meet the user requirements and are late, over budget, and of poor quality [Standish 2020]. A 2002 study from the National Institute of Standards and Technology [Tassey 2002] reports that software defects are so prevalent and so detrimental that they cost the U.S. economy an estimated $59.5 billion annually, or about 0.6% of the gross domestic product. Parnas and Lawford [Parnas 2003] state "Despite more than 30 years' effort to improve software quality, companies still release programs containing numerous errors. Many major products have thousands of bugs".

Although, software development has improved considerably in the last 40 years, the development and maintenance of software systems are still plagued by serious budget, schedule and quality problems. One of the reasons for this is the lack of software engineers with adequate qualifications and background. There is a very high demand for software engineers; the *Occupational Outlook Handbook 2016–17 edition* (Bureau of Labor Statistics) [BLS 2016] highlights the increased need, in the coming decade, for software engineers. The handbook states, "Employment of software developers is projected to grow 17% from 2014 to 2024, much faster than the average for all occupations. The main reason for the rapid growth is a large increase in the demand for computer software".

THE NATURE OF SOFTWARE

Another reason for the problems in software engineering is that by its nature, software is different from other engineered products. In the paper "No silver Bullet" [Brooks 1995], Brooks argues that software has the following "essential difficulties" that do not plague other engineering products (or at least not to the same extent):

- *Complexity* – Brooks argues that software entities are more complex, for their size, than perhaps any other human construct. Even moderately complex software systems may have millions of distinct parts (components/packages, classes/objects, variables, methods, and statements), which may be stored in an extremely small physical space.
- *Conformity* – Other engineering fields (such as mechanical engineering) deal with complexity by building models that conform to the known laws of physical world (e.g., Newton's laws of motion). Software engineers also build models, but software does not have the unifying principles based on the physical world that it must conform to; hence, software models must deal with an arbitrary conformity imposed by the expected software capability – often rather open-ended, depending on user/customer needs.
- *Changeability* – Software is easy to change, at least relative to changing a physical object. For example, adding a turn arrow to the traffic signals in a city's traffic system is a major undertaking, while making such a change to a simulated traffic system might be the weekend activity of a single programmer. But the ease of software change has its negative side. Changing software in haste can lead to injecting defects into a system or can have unintended side effects in some other part of the system (a common occurrence). Also, the changeability nature of software can lead users/customers to expect quick and cheap addition of features to a software product.
- *Invisibility* – Software is invisible. It lacks physical structure and substance. Although software engineers use graphic models to help them to understand software's nature (e.g., a use case diagram – see Chapter 5), such graphic models are abstractions and only represent some aspect of the software's use or organization, not its "physical appearance".

ADVANCING SOFTWARE ENGINEERING

In the past decade, there has been significant progress in the advancement of software engineering as a profession. In recent years, task forces and working groups from the Association of Computing Machinery (ACM), the Institute of Electrical and Electronics Engineers Computer Society (IEEE-CS), the Accreditation Board for Engineering and Technology (ABET), and the Software Engineering Institute (SEI), at Carnegie Mellon University, have been at work to advance the state of software engineering, and have developed the following components, which help define and characterize the nature and content of software engineering and to improve its practice:

- Accreditation Criteria for Software Engineering Programs [ABET 2018]
- Software Engineering Code of Ethics and Professional Practice [ACM 1999]
- Undergraduate and Graduate Reference Curricula [ACM 2009, 2015]

- Certified Software Development Professional (http://www.computer. org/certification/)
- CMMI (Capability Maturity Model Integration) for Development [SEI 2010]
- Guide to Software Engineering Body of Knowledge [Bourque 2014]
- A Software Engineering Competency Model (SWECOM) [SWECOM 2014]

THE SOFTWARE ENGINEERING BODY OF KNOWLEDGE (SWEBOK)

The SWEBOK [Bourque 2014] has had a major influence on understanding the nature of software engineering and on the development of education and professional development programs. It is also a major influence on the content and character of this book. The SWEBOK provides the description of the Software Engineering Practice Knowledge Areas (KAs) listed in Table 1.1. Topics from each of the "Practice" KAs will be covered in this text, which is referred to in the Location column of Table 1.1. The SWEBOK also includes KAs labeled as Software Engineering Foundations: Engineering Economy Foundations, Computing Foundations, Mathematical Foundations, and Engineering Foundations.

Table 1.2 describes some of the different roles and responsibilities that a software engineer might take on. This arrangement of software engineering roles will be used in composing the *DigitalHome* project team with each member also assuming the role of a component developer. The roles in an agile team are discussed in Chapter 10.

Table 1.1 SWEBOK Knowledge Areas for Software Engineering Practice

Knowledge Area	Description	Location
Software Requirements	Concerned with determining, documenting, and managing the requirements for a software product or service.	Chapters 5,10
Software Design	Focused on deciding how the software requirements will be fulfilled: determining the design components and their description, and how they will interact with each other.	Chapter 6

(Continued)

Table 1.1 (*Continued*)   SWEBOK Knowledge Areas for Software Engineering Practice

Knowledge Area	Description	Location
Software Construction	Concentrated on the design, coding, and testing the units that make up the software design components.	Chapter 7
Software Testing	Ensuring through tests on an executing software system or its units/components that the software operates properly.	Chapter 3
Software Maintenance	Covers the evolution of software as it changes over its operational life.	Chapter 8
Software Configuration Management	Encompasses the management of the software artifacts (such as plans requirements, design, and code) of software as it is developed: version control, change control, and document archiving.	Chapter 4
Software Engineering Management	Involves planning, coordinating, measuring, reporting, and controlling a software project.	Chapter 4
Software Engineering Process	Includes the definition, implementation, evaluation, and improvement of the processes used to develop software.	Chapters 2, 10
Software Engineering Models & Methods	Describes models and methods used to develop and maintain software.	Chapters 2, 8, 10
Software Quality	Covers the processes and practices used to assure software quality.	Chapters 3, 10

(*Continued*)

Table 1.1 (*Continued*)   SWEBOK Knowledge Areas for Software Engineering Practice

Knowledge Area	Description	Location
Software Engineering Professional Practice	Concerned with the capabilities and attitudes that software engineers must possess for professional, responsible, and ethical practice of their profession.	Chapter 9

Table 1.2   *DigitalHome* Development Team's Role Description

Role	Description
Team Leader	The team leader is responsible for guiding and motivating team members throughout the development phases. This includes running the day-to-day operations, resolving conflicts, managing and tracking issues, conducting meetings, and managing team resources. He/she serves as the team's interface with the customer and management. The team leader is also responsible for monitoring team members' work and ensuring process discipline. An effective team leader strives for open and effective team communication. He/she makes sure that project's status is communicated to the rest of the team members as well as to management and the customer.
Software Analyst	The software analyst is responsible for identifying the purpose of the software system and the individual goals of the customer. He/she leads the development of the requirements specification that includes a model of the system's environment and a description of the purposes, goals, performance constraints, resource constraints, implementation constraints and external interfaces of the system. The analyst must know the technology and be able to understand and respond to what is found in observing and talking with those who are commissioning a new system or will be the end users of it. The software analyst must have considerable written communication as well as generalization skills.

(Continued)

Table 1.2 (*Continued*) *DigitalHome* Development Team's Role Description

Role	Description
Software Architect	The software architect is concerned with the structure of the software system. In particular, an architect leads the team in defining the software components and the interactions among these components with respect to the specification. He/she leads the team in the development of the high-level architecture as well as the high-level and detailed-level component designs. The architect is responsible for recording decisions that determine the implementation strategy. He/she must have underlying technical knowledge of component development (modular design, programming, and testing) in order to tie his/her decisions to reality.
Quality Manager	The quality manager leads the team in producing and tracking the team quality plans, as well as alerting the team leader to arising quality issues. The team quality manager assures that updates to quality plans are done periodically and accurately. He/she is also responsible for making sure that Verification and Validation activities (reviews, inspections, tests) are performed properly and that the results of these activities are recorded correctly and appropriate changes and follow-ups are completed. This person is also responsible for making quality improvement recommendations at the end of each development cycle to assure continuous quality improvement of process and product.
Planning Manager	The planning manager leads the team in producing a detailed, balanced, and accurate team development plan as well as individual plans. He/she is responsible for tracking progress against developed plans, and alerts the team leader in case of deviation from these plans. This manager ensures that team members measure and record their work (effort and size). The planning manager leads the team in producing the team schedule for each development cycle, as well as producing a cycle report at the end of each cycle.

(Continued)

Table 1.2 (*Continued*)   *DigitalHome* Development Team's Role Description

Role	Description
Component Developer	The module developer is responsible for developing and integrating the modules that make up an architectural component. He/she develops the detailed design for the modules, translates the design into program code, tests the modules, and integrates them into a component.

Forming a Project Team

At the 202X Homeowner Retreat, Jose Ortiz was appointed the head of the new *DigitalHomeOwner* Division of *HomeOwner*. One of his first tasks was to form a project team. The following sections describe how the team was formed and discusses the team members' background and experience.

Jose Ortiz was born in San Diego, CA, in 1970. He received a B.S. in electrical engineering at San Diego State University and subsequently earned an MBA at California State at Transylvania. After working at several aerospace firms in the 1990s, Jose was employed as a technical manager at a small IT start-up firm, Network Solutions, in 1999. After the firm's IPO (Initial Public Offering) in 2000 it had tremendous growth and success. In 2010, Jose was hired by HomeOwner as its Deputy CIO.

In early September 202X, Jose began to assemble a software team to develop the *DigitalHome* prototype. Although Jose had outstanding management skills and excellent technical knowledge about network administration, IS evolution, and IT architectures, he had little detailed knowledge about software development and recent advancements in software engineering methods and practices.

To help him in the team formation task, Jose contacted his old friend **Michel Jackson**, who retired after a 30-year career in software development. As a new Stanford graduate in Computer Science in 1982, Michel started his career as embedded systems programmer for a major aerospace company. Since then Michel worked in almost every facet of the software business: as a programmer, a software architect, a software/system analyst, a quality assurance director, a project manager, and an entrepreneur. He has worked on large, custom built real-time, embedded systems, shrink-wrapped software, and major web-centric applications. Most recently, Michel headed a small company that specialized in software analysis and modeling, exploiting the use of the Jackson Analysis Framework. After selling his firm last year he has been in quasi-retirement, with part-time consulting work occupying his non-leisure time.

Michel eagerly agreed to help Jose in recruiting a *DigitalHome* development team. Jose and Michel conferred about the team organization for the project and agreed that the team should be organized in a quasi-democratic manner, with each member assigned prescribed responsibilities and all major decisions to be made after team discussion, with the Team Leader making the final call. Table 1.2 contains the description of the DH Team roles. All team members may be assigned responsibility for a software unit (component/unit design, coding, and testing).

Michel first contacted **Disha Chandra**, who he had met when he did some consulting work for *SoftMedic*. Disha had come to *HomeOwner* three years ago, after a decade of work at *SoftMedic*. At *SoftMedic* she worked in a variety of roles on the development and maintenance of a number of widely used software applications in the field of health delivery and management. Most recently, at *Homeowner*, Disha lead the development of an in house application used by *HomeOwner* store managers to train new employees.

Disha is 38 years old and came to the U.S. from India in 2000 to work on a degree in Computer Science at Valley State University. After completing her degree, she went to work for *SoftMedic*, became a U.S. citizen, and earned an online Master of Software Engineering degree from Lancaster Institute of Technology. She is considered a strong leader and an effective manager with excellent knowledge and experience in almost all areas of software engineering.

Jose interviewed Disha and was very impressed with her experience, and the confident and organized way she approached her work. He offered her the job of Team Leader. She accepted with one stipulation, she wanted Michel to serve as the project's Software Analyst. Jose and Disha conferred with Michel about the Software Analyst role and Michel was enthusiastic about working on the project and accepted the offer.

From this point, Disha and Michel worked together to form the rest of the team. In the next ten days they contacted and interviewed eight persons for the *DigitalHome* project. After some consideration and agreement from Jose Ortiz, the following team members were selected.

- **Yao Wang**, Software Architect, DH System
  Yao Wang is considered a whiz kid of object-oriented design. At 26 years of age he shows great promise to be an innovative leader in software design of embedded consumer products. He was hired by Jose Ortiz right after graduation from Rational University and is viewed as future principal player in the *DigitalHomeOwner* Division.
- **Georgia Magee**, Quality Manager, DH System
  Georgia Magee has worked for the last three years as a programmer and a test engineer at *HomeOwner*. Prior to that, she had a four-year stint with the Volcanic Power Company, as a junior software engineer developing electric power management software. Georgia is married, 30 years old, and recently became a mother with the birth of her first child, George.

■ **Massood Zewail**, Planning Manager, DH System

Massood recently graduated from the University of Central California (UCC) with a joint degree: an undergraduate degree in computer engineering and master's degree in software engineering. Although Massood was an outstanding student (3.87 GPA) and had two summers of student intern work with MacroSoft Corporation working on two different projects which followed two different development processes (spiral and agile), this is his first full-time professional employment.

As part of the team formation, Disha, in consultation with the other members, decided that each team member would have additional responsibility as a Component Developer, developing one or more software components.

EXERCISE 1.1: TEAM PROBLEMS

During the project start-up, Jose Ortiz wanted to build a strong team that would work well together. Since none of the team members had worked together before, he designed a team activity where team members would discuss teamwork problems that they had experienced in previous projects.

The exercise, described at the end of this chapter, presents some of the past problems experienced by the DH Team members and asks questions about resolving the problems.

Assessing *DigitalHome's* Needs

One of the key decisions at the August 202X *HomeOwner* retreat was for *DigitalHomeOwner*, with the assistance of the Marketing Division, to conduct a needs assessment for a *DigitalHome* product that would best meet the needs and desires of homeowners. In early September, Jose Ortiz met with Dick Punch, VP of Marketing, and they decided on a set of parameters to use in developing a Need Statement for the *DigitalHome* product. Dick agreed to have the Marketing Division draft two documents in the coming weeks: a *DH Customer Need Statement* and a *DH High-Level Requirements Definition* (HLRD).

DH Customer Need Statement

The Marketing Division of *HomeOwner* produced a draft Need Statement for DH and submitted it to Jose Ortiz for review and feedback. The Need Statement was based on research that involved structured interviews and focus groups of potential DH customers and users. The results were analyzed and assembled into statements about the needs of a fictional Wright family, who represent a set of typical potential DH users. The *DH Customer Need Statement* is in Appendix A.

DH High-Level Requirements Definition

After some discussion about the *DH Need Statement*, the Marketing Division prepared a *DH High-Level Requirement Definition (HLRD)*. The HLRD is not meant to cover all the expectations in the Need Statement, but rather represents a subset of needs that can form the basis for a prototype development. As such, it follows the directive issued in the *HomeOwner* August retreat: "Choose a set of features for the *DigitalHome* product to develop as a prototype to illustrate how the *DigitalHome* needs will be met. Because of time constraints, the features chosen will be a subset of the actual needs". The HLRD is included in the below section.

HIGH-LEVEL REQUIREMENTS DEFINITION

INTRODUCTION

This is a "high-level" definition of the requirements (HLRD) for the development of a "Smart House", called *DigitalHome*, by the *DigitalHomeOwner* Division of HomeOwner Inc. A "Smart House" is a home management system that allows home owners (or renters) to easily manage their daily lives by providing for a lifestyle that brings together security, environmental and energy management (temperature, humidity and lighting), entertainment, and communications. The Smart House components consist of household devices (e.g., a power and lighting system, an air conditioning unit, a sound system, a water sprinkler system, small appliances, security system, etc.), sensors and controllers for the devices, communication links between the components, and a computer system, which will manage the components.

The HLRD is based on the *DigitalHome Customer Need Statement*. It is made up of a list of the principal features of the system. This initial version of *DigitalHome* will be a limited prototype version, which will be used by HomeOwner management to make business decisions about the future commercial development of *DigitalHomeOwner* products and services. Hence, it does not include all features discussed in the *Need Statement,* and the HLRD is not intended as a comprehensive or complete specification of *DigitalHome* requirements. The main purpose of the HLRD is to support an effective project planning activity. The HLRD was prepared by the Marketing Division of HomeOwner Inc, as part of a needs assessment for the *DigitalHome* project.

DIGITALHOME PROTOTYPE FEATURES

1. The *DigitalHome* System shall allow any web-ready computer, cell phone, or other device to control a home's temperature, humidity, lights, and the state of household appliances (coffee maker, stove, etc.). The communication center of the system shall be a personal home owner web page, through which a user can monitor and control home devices and systems.

2. Each *DigitalHome* shall contain a master control device that connects to the home's broadband Internet connection, and uses wireless communication to send and receive communication between the *DigitalHome* system and the home devices and systems.

3. The *DigitalHome* shall be equipped with various environmental sensors (temperature sensors, light sensors, humidity sensors, power sensors, contact sensors, water sensors, etc.). Using wireless communication, sensor values can be read and saved in the home database.

4. The *DigitalHome* security system consists of a set of contact sensors and a set of security alarms, which are activated when there is a security breach.

    a. The security system shall use wireless signals to communicate through the master control unit.

    b. The system shall use both sound and light alarms and will be able to manage up to 30 door and window sensors.

5. The *DigitalHome* programmable Thermostat shall allow a user to easily monitor and control a home's temperature from anywhere using any web-ready computer, cell phone, or other device.

    a. Thermostats can be placed throughout the home and can be controlled individually or collectively, so that temperature can be controlled at different levels in different home spaces.

    b. A thermostat unit shall communicate, through wireless signals, with the master control unit.

  c. The system shall support Fahrenheit and Celsius temperature values.

  d. The system shall be compatible with most centralized HVAC (Heating, Ventilation and Air Conditioning) systems: gas, oil, electricity, solar, or a combination of two or more.

  e. The user shall always be able to override the scheduled settings at any time.

6. The *DigitalHome* programmable Humidistat shall allow a user to easily monitor and control a home's humidity from anywhere using almost any web-ready computer, cell phone, or other device.

  a. Humidistats can be placed throughout the home and can be controlled individually or collectively, so that humidity can be controlled at different levels in different home spaces.

  b. A Humidistat unit shall communicate, with wireless signals, through the master control unit.

  c. A Humidistat unit shall manage humidity sensors and dehumidifiers/humidifiers located in a specified home space.

  d. The user shall be able to select the humidity levels found most comfortable — from 30% to 60%.

7. The *DigitalHome* programmable Power Switch shall provide for management of a home's household appliances and shall allow the user to turn appliances and lights on or off as desired.

  a. The Power Switch unit can control the central lighting in each room and up to forty 115-V, 10 amp appliances that plug into a standard wall outlet.

  b. The system shall be able to provide information about whether an appliance or a light is off/on.

  c. A user shall be able to monitor the state of the appliance, and turn on or off any appliance through any web-ready computer, cell phone, or other device.

  d. The sensors should be able to monitor the state of the appliance and manipulate the power setting in order to reach the highest energy efficiency. For example, monitor the water temperature and adjust water heater power supply accordingly to increase its energy efficiency.

8. The *DigitalHome* Planner shall be able to provide a user with the capability to direct the system to set various home parameters (temperature, humidity, security level, and on/off appliance/light status) for specified time periods.

  a. *DigitalHome* provides a monthly planner on its website.

  b. Parameter values can be scheduled on a daily or hourly basis.

   c. All planned parameter values can be overridden by a user.
   d. Various plan profiles (normal monthly profile, vacation profile, summer profile, holiday profile, etc.) may be stored and retrieved to assist in planning.
   e. The *DigitalHome* Planner shall be able to provide various reports on it management and control of the home (e.g., historical data on temperature, humidity, lighting, etc.).

Dick Punch told Jose Ortiz that he understood that it was typical at the beginning part of a project to determine a "concept of operations". Jose agreed, and Dick asked if Jose could prepare a presentation of the concept.

MINITUTORIAL 1.2: CONCEPT OF OPERATIONS

This HLRD (along with the *DH Customer Need Statement* and the DH Conceptual Design in the next chapter) could be considered a Concept of Operations (ConOps) document. Appendix B in the IEEE standard life cycle processes — Requirements engineering [IEEE 29148] describes a ConOps as "the organization's assumptions or intent in regard to an overall operation or series of operations of the business with using the system to be developed, existing systems, and possible future systems". The ConOps document can be used to communicate overall system characteristics (both quantitative and qualitative) to the user, customer, developer, and other organizational entities (e.g., training or maintenance staff). We note that for the DH System we are speaking of a "software-intensive" system.

The ConOps Document for complex, multifaceted systems would provide the following:

■ A description of the elements of the user organization.
■ A description of the user's operational needs without including detailed technical issues, which would be addressed in subsequent development activities (e.g., requirements analysis).
■ Documentation of system characteristics that can be verified by the user without requiring any special technical knowledge beyond normal user understanding.
■ A place for users to state their desires, visions, and expectations.
■ A mechanism for users and customers to express thoughts and concerns on possible solution strategies.

The IEEE standard [IEEE 29148] provides a detailed outline for a ConOps document, addressing each of the above bulleted features. The DH Team did not feel that a full ConOps document, with all of the above features, was needed or appropriate for the DH System: the DH System has a prototype focus, and the primary user group for DH is an amorphous, unstructured group.

EXERCISE 1.2: ASSESSING CUSTOMER NEEDS

Disha Chandra set up a meeting between the DH Team with Karen Mullen, the lead for the DH needs assessment effort, to address questions and concerns about the *DH Customer Need Statement* and the *DH HLRD*.

The exercise, described at the end of this chapter, deals with preparation for the meeting.

Case Study Exercises

EXERCISE 1.1: TEAM PROBLEMS

During the *DigitalHome* project launch, Jose Ortiz wanted to build a strong team that would work well together. Since none of the team members had worked together in the past, he designed a team activity where team members would discuss teamwork problems that they had experienced in previous projects.

This exercise presents some of the past problems experienced by the DH team members and asks questions about resolving the problems.

LEARNING OBJECTIVES

Upon completion of this exercise, students will have increased ability to:

- Describe some of the personnel problems that may be experienced by software project teams.
- Develop solutions to team personnel problems.
- Appreciate the need for project team members to work together effectively.

EXERCISE DESCRIPTION

1. As preparation for this exercise, read the Chapter 1 material on the DH Team Formation.
2. You will be assigned to a small team. Each team is assigned one of the team problem descriptions listed below, which describes a team personnel problem.
3. The team meets and discusses the problem, and answers the questions about the problem. Then, the team decides on various approaches that might be taken to solve the problem.
4. Each team summarizes its discussion and conclusions in a team report. Choose one member of their team to report to the class on the team's discussion and conclusions.

GEORGIA MAGEE'S PROBLEM: DON'T WORRY YOUR PRETTY LITTLE HEAD

While working at Volcanic Power Company, Georgia served in a project team to develop a power regulation product. One of Georgia's teammates was a young software engineer named Kevin: Kevin had been writing programs since he was in middle school, had graduated with honors in computer science from Carnegie Mellon University, and was considered a real whiz at constructing software. The Team Leader, Roy, is an easygoing fellow who just wants to get through the project without making too many tough decisions. During project launch, Kevin argued strongly that he was best qualified to be the development manager (responsible for managing system analysis and design). Without consulting with anyone else, Roy designated Kevin as the project Development Manager. So far, in the first development cycle, Kevin has completed the whole design by himself (without consulting with anyone) and written nearly all of the source code. The team is ready to go into system test and Kevin has told everyone not to worry, he is great at testing software and he will handle it. Georgia was assigned the role of Process/Quality Manager. She complained to Roy that because Kevin is doing everything, the team is not really following the Company's established development process. Roy tries to convince Georgia that things are going great, Kevin is doing a great job, and not to worry her pretty little head about such things as "design" and "coding" and just concentrate on the process stuff.

Suppose you are the VP of Information Technology at Volcanic Power and Georgia brings her complaints to you. What would you do?

MASSOOD'S PROBLEM: SLACKER ON THE LOOSE

While attending the UCC Massood was enrolled in a senior design project course, which involved development of an autonomous chemical sensing robot. The design team was made up of computer science and computer engineering majors. One of the team members, Burt, has spent more time in his life trying to get out of work than actually working. He seems to really enjoy concocting schemes for avoiding labor of any sort. He is bright enough, and this has allowed him to stumble through three years of a CS program at UCC with a C minus average. Unfortunately, he is now enrolled in the senior design course and has been assigned as a programmer on the project team. Each week, Burt is supposed to submit a report on his work (tasks completed, time spent on the project, and a description of any problems incurred) to Massood, who is the project Planning Manager. Massood became really fed up with Burt. Burt's reports were always late and incomplete, and they look like they were made up minutes before submission. Massood discussed this with the other team members, and they all agree that Burt was a slacker and had contributed next to nothing to the project. They urged the Team Leader, Juan, to discuss this with the teacher. They would like to get Burt off their team before he does too much damage.

Juan met with you, the teacher, explained the situation, and asked that Burt be removed from their team. If you were the teacher what would you do?

DISHA'S PROBLEM: IT'S TIME TO MAKE SOME CHANGES

On her first project at *SoftMedic,* Disha was a member of a team for which Sally was the Team Leader. Sally was intelligent, hard-working, and very ambitious; but she had never led a project of this magnitude. After the first cycle, the *SoftMedic* Quality Assurance Director gave the team an overall assessment of the team's work. Most would consider Sally's team "average". They worked fairly hard, but were a technology-oriented group and did not aspire to move up the corporate ladder, like Sally. Sally wanted the team to be viewed as outstanding and made this clear to her teammates. In the second cycle, Sally has taken a firmer hand and has now given direct orders to each of the team members about how and when they should do their work. If work is not completed the way she expects, she will berate a team member in the weekly team meeting. After the cycle 2 design inspection, Sally stated

> It's clear to me that you guys just don't care about this project, and I think it is time to make some changes! From now on, we will meet every evening at 6, and I expect each of you to make a daily report on what you have been doing.

The team, mostly a mild-mannered group, does not dispute Sally's proclamation and walks off mumbling to themselves.

As the VP of Development, one of the team members comes to you to complain about the situation. What should you tell the team member? What should you do with Sally, if anything? How should you handle the situation?

Yao's Problem: Double Trouble

Yao's last project at HomeOwner involved development of an inventory application that could be embedded in checkout scanners and handheld devices. Yao was designated as the Design Manager and he looked forward to the project. However, two of the team members caused problems: Greg and Whitney just do not like each other; they had been involved in several run-ins in the past. They were both excellent engineers, but just could not get along. Greg was the Requirements Analyst and Whitney the Planning Manager. The team had just completed the requirements phase and was starting the design phase. During the requirements inspection meeting, Greg and Whitney had a big disagreement about the way one of the requirements was written, and almost came to blows. In the first meeting about the design, Yao presented a preliminary design architecture. Greg and Whitney proceeded to spend almost the entire meeting bickering over every detail of the architecture. Finally, Harry, the Team Leader, said "Enough! We can't do anything productive until you two stop acting like children". After the meeting Harry met with both of them, separately, and tried to resolve the problem, but he could not make any headway.

Harry comes to you, the Division Head, for help. What do you do?

EXERCISE 1.2: ASSESSING CUSTOMER NEEDS

The *HomeOwner* Marketing Division produced the two documents: the *DH Customer Need Statement* and the *DH HLRD*. After reviewing the documents with her team, Disha Chandra asked Jose Ortiz if he could set up a meeting between the Marketing Division and the DH Team to discuss the questions the team had about the Need Statement and the HLRD. Jose contacted the Marketing VP and set up a meeting between the DH Team and Karen Mullen, the lead for the DH needs assessment effort. This exercise deals with how the team will prepare for the meeting with Karen.

Learning Objectives

Upon completion of this exercise, students will have increased ability to:

- Analyze a customer need statement and the initial set of requirements for a system.
- Acquire additional information from a customer about his/her needs.
- Work effectively as part of a team.
- Prepare and organize for a meeting.

Exercise Description

1. As preparation for this exercise, read Chapter 1.
2. You will be assigned to a small development team (like the DH Team).
3. Your team is to take on the role of the DH Team and prepare for a meeting with Karen Mullen. The team should carry out the following tasks:
   a. Analyze the DH HLRD and discuss any problems or concerns about their understanding of the HLRD.

b. Formulate objectives for the meeting with Karen.
c. Make up a set of questions the team would like answers to prior to commencement of project planning and software requirements analysis.
d. Assign individual roles for the meeting (e.g., meeting facilitator, taking notes, asking questions, etc.).
e. Make up an agenda for the meeting.
f. Complete the following meeting preparation form.

Meeting Preparation Form	
Meeting Date & Time	
Meeting Purpose	
Meeting Participants and their Roles	
Meeting Agenda	
Questions 1	
2	
3	
4	
5	
6	
7	
8	
9	
10	

Chapter 2

Launching *DigitalHome*

Project Launch

On 9/15/202X, the DH Team held a meeting with Karen Mullen, the lead for the DH needs assessment effort to discuss questions the team had about the *DH Customer Need Statement* and the *DH High-Level Requirements Definition* (HLRD), prepared by the *HomeOwner* Marketing Division. After the meeting, the HLRD was finalized.

In mid-September 202X, a two-day workshop meeting was held at the conference room of the office of the *DigitalHomeOwner* Division of *HomeOwner* to launch the *DigitalHome* project. In attendance was the Director of the *DigitalHomeOwner* Division, Jose Ortiz, Mr. Ortiz's Admin Assistant, Stella Washington, the DH development team leader, Disha Chandra, and the other team members: Michel Jackson, Yao Wang, Georgia Magee, and Massood Zewail. Jose and Disha had met earlier and decided on the agenda for the meeting and developed a Project

Launch Process (see Table 2.1). Part of the launch had already been completed: team roles and responsibilities had been assigned; and the team had studied the Need Statement and the HLRD and had met with a *HomeOwner* marketing representative to better understand what needed to be accomplished (see Chapter 1).

Table 2.1   DigitalHome Launch Process Script – 9/16/202X

Purpose	• Guide the *DigitalHome* project launch
Entry Criteria	• DH Customer Need Statement • DH HLRD
Activity	*Description*
Team Formation	• Determine and assign team roles and responsibilities • Decide on individual and team goals • Establish communication and reporting methods and protocols • Determine support infrastructure (staff, tools, and facilities) • Establish the DH Development Process • Determine initial project budget
Need Analysis	• Study and analyze the Need Statement and HLRD • Hold meeting with *HomeOwner* marketing representatives to assist in needs analysis • Write Needs Analysis Report
Conceptual Design	• Develop a *DigitalHome* Context Diagram • Develop a *DigitalHome* Conceptual Design
Development Strategy	• Determine criteria for a development strategy • Create Development Strategy • Determine the development process • Determine the number of development cycles • Allocate development modules in each cycle • Estimate module size and development effort for each cycle
Postmortem	• Perform postmortem analysis of project launch activities
Exit Criteria	• Needs Analysis Report • Development process has been chosen • Team formation has been completed • DH Context Diagram • DH Conceptual Design • DH Development Strategy

Team Building

After the two-day workshop, Jose Ortiz and Disha Chandra met and decided to deliver a mini-tutorial on team building.

MINITUTORIAL 2.1: BUILDING AN EFFECTIVE TEAM

Building a successful project team is one of the most crucial elements influencing project success. When project teams fail to exhibit effective teamwork, the project is more likely to have problems with schedule, budget, and quality. There is extensive research and analysis on the characteristics of effective teams and how to build them [DeMarco 1999, Ganis 2007, Humphrey 2000, Oakley 2004].

The following positive qualities have been found to be characteristic of effective teams:

- Team members function efficiently and productively.
- The teams share leadership and accountability for their work products.
- Emphasis is shifted from the individual to the team.
- Individual performance is based on achieving team goals.
- Team operates in an environment of open and honest communication.
- Problems are discussed and resolved by the team.
- The team as a whole is greater than the sum of its parts.

The following problems often characterize ineffective teams:

- Poor communication within the team
- Inability to compromise or cooperate
- Weak participation by a team member
- Some team members do the majority of the work
- Lack of discipline
- Unclear decision-making process
- Ineffective leadership

The question arises of how we build an effective team; that is, how do we form a team that possesses the above listed positive qualities and not the problems of ineffective teams. Unfortunately, many managers fail to appreciate the effort that has to be invested and the attention to detail needed for successful team formation. Forming an effective team is not just a matter of throwing a set of capable individuals together to accomplish a common goal (often an ill-defined goal).

There are lots of ideas and opinions, some divergent, on how to build effective teams. The following discussion presents techniques that have proved the most fruitful.

SET GOALS

Setting team goals provides the foundation for successful teamwork. This can provide a shared sense of purpose, which unifies the team members and offers a framework for deciding on team direction and for assessing team performance. Goals should be measurable in order to assess project progress and to provide guidance on how to make adjustments, when there are problems. This means that progress toward reaching goals should be regularly monitored and measured.

The following are some example team goals:

- Milestones, and associated deliverables, will be completed as scheduled.
- The team will follow the prescribed changed control process.
- Decisions will be made in a collaborative manner.
- Data will be collected for the project (development process, productivity, quality customer satisfaction, etc.) and analyzed to assess progress, and product, and process quality.

DEFINE ROLES AND RESPONSIBILITIES

It is critical that each team member understands her/his role on the team and there is a clear description of the responsibilities and expectations for that role. This makes it easier to assign task leadership and establish solid working relationships between team members and to assess individual performance. It is unrealistic to expect all members of a team to undertake appropriate roles and responsibilities without some guidance and direction.

Table 1.2 in Chapter 1 provides an example of how roles and responsibilities might be defined. Notice that each role description includes a responsibility for leading or managing a significant part of the project activities. This is an example of shared responsibility, which is critical for the team to feel that work is distributed equitably and that all team members are essential to team success. It also illustrates that a variety of talents and capabilities are needed in such projects.

ESTABLISH COMMUNICATION AND DECISION-MAKING PROCEDURES

Even though team members may have reasonably good oral and written communication skills, in a common language, there are a number of issues that might negatively affect good communication:

■ One's cultural background might influence how others understand the meaning and significance of what is being communicated.
■ Team members do not use active listening as part of their interaction with each other.
■ The team is too large or poorly organized for one-on-one or small group communication.
■ The team decision-making structure discourages or obviates open communication.
■ There are inadequate defined procedures and methods for communication: insufficient documentation standards, regular team meetings, and status reports; and limited use of communication technology such as online discussion groups, resources for storing and exchanging team documents, and the use of electronic communication.

An ill-defined or ineffective decision-making procedure can doom a team to failure. There are various team structures that influence communication patterns among members and prescribe the decision-making paradigm:

■ *Controlled Centralized* – this is top-down hierarchical decision structure, much like that existing in military organizations.
■ *Democratic Decentralized* – all decisions are made by group consensus of the team; all work is considered group work.

■ *Controlled Decentralized* – attempts to combine the benefits of centralized and decentralized control; there is centralized decision team, with communication decentralized to the rest or team (or subteams if the overall team is very large).

The team structure must fit the organizational structure, the application domain, the size and complexity of the project, and the competencies and experience of the team members.

ENSURE A TEAM HAS CONTROL OF ITS WORK

Effective teams must have some control over their work, the more the better. While complying with project constraints (e.g., budget and schedule constraints), teams should control how they organize and plan their work and how they go about their day-to-day tasks. This approach supports building cohesive, committed teams.

START WITH TEAM BUILDING ACTIVITIES

Team building activities can be conducted by a "team building" trainer or by a team member who has had success in building and leading effective teams. However, even seasoned professionals may need training – to be introduced to or reminded of what it takes to build an effective team. Examples of such activities include the following:

■ Develop a set of Team Goals
■ Complete the Team Problems exercise in Chapter 1
■ Develop a Team Policies Statement (describes team roles and the responsibilities, defines the decision-making policy, presents strategies for dealing with disputes between team members, describes the procedures for holding a team meeting – defined agenda, rules for interacting, meeting roles)
■ Develop a Team Expectations Agreement (a common set of realistic expectations that the members generate and agree to honor)

EXERCISE 2.1: FORMING AN EFFECTIVE TEAM

Jose Ortiz and Disha Chandra felt after the team building tutorial the team needed to start working on forming a cohesive and effective team. The earlier Team Problems activity (see Chapter 1) was a good start, but they both felt an additional Team Formation activity was needed.

The exercise, described at the end of this chapter, deals with the Team Formation activity designed by Jose and Disha to help the DH Team become an effective cohesive working group.

Software Development Process

MINITUTORIAL 2.2: SOFTWARE PROCESS FUNDAMENTALS

A process is a set of practices performed to achieve a given purpose. A *defined process* would include such things as a process script, tools, methods, standards, and identified roles and responsibilities. Processes differ according to project scope, complexity, and domain; and they must accommodate a business' goals and culture, and the developer experience. While process is often described as a leg of the *process-people-technology* triad, it may also be considered the "glue" that unifies the other aspects.

Software processes are designed and used to improve the development of software – cost, schedule, and quality. Table 2.1 displays a process script that describes a process for launching a software project (although it certainly could be used for non-software projects).

PLAN-DRIVEN PROCESSES AND AGILE PROCESSES

There are many approaches and opinions about how software should be developed. Boehm and Turner [Boehm 2004] provide an excellent summary and overview of various software methods and approaches. Processes are often classified as "Plan-Driven" or "Agile".

Plan-Driven Processes (sometimes called "heavyweight" or "traditional") typically incorporate repeatability and predictability, a defined process, extensive documentation, up-front system architecture, detailed plans, process monitoring and control, risk management, software configuration management, and well-formulated verification and validation. They are often criticized for their inflexibility, lack of innovation, and poor response to the need for change. Advocates of plan-driven methods point to the need for structure and discipline in the development of large, complex systems with long development schedules, large developer teams, and mission-critical systems dependent on safety, security, reliability, or other quality attributes that require a rigorous process.

Agile Processes (sometimes call "lightweight") have been developed to address the criticism that plan-driven processes are unresponsive to change (e.g., change in requirements or change in technology). Agile processes have the goal of reducing the cost and time to make changes throughout a project. Agile developers advocate the following (referred to as the "Agile Manifesto") [Highsmith 2001]:

- Individuals and interactions over processes and tools
- Working software over comprehensive software documentation
- Customer collaboration over contract negotiation
- Responding to change over following a plan

As with most two-category classification systems, the plan-driven processes versus agile processes debate often captures the extremes and ignores the middle ground. As we will see in the following discussion of various software processes, there is a "middle-weight" division – that is, in the application of any process model some combination and agile and plan-driven methods may be used.

SOFTWARE LIFE CYCLE MODELS

A software development life cycle (SDLC) consists of the various phases and activities that a software product or service goes through during its "life" – from its initial conceptual formulation to its specification, design, construction, and test, and forward to its operation and maintenance, and finally to its retirement/disposal.

A software life cycle model is a high-level depiction of the life cycle phases and how they relate to each other. In this mini-tutorial, we do not discuss all such models, but provide an overview of several popular models with intent to show the variety and flexibility of models. In the next section, we describe in more detail specific processes used to develop software.

Waterfall Model

The waterfall model is a SDLC that is intended to progress through its life cycle phases in a sequential manner, as depicted Figure 2.1. In reality, this is rarely the

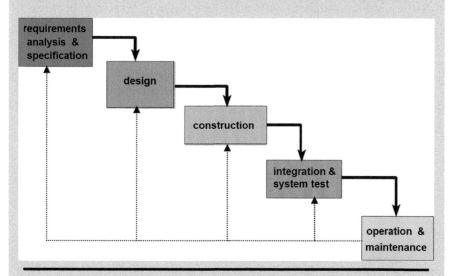

Figure 2.1 Waterfall model.

case. For example, if in the design phase a problem is discovered in the statement of a software requirement (maybe an imprecise or ambiguous statement), it may be necessary to return to the Requirements Analysis and Specification phase. Such situations occur often, especially in the development of large, complex systems. So although a linear or sequential SDLC may be intended, it is typical in practice that the process is iterative – cycling back through previous phases as problems are encountered or there are changes mandated by circumstances out of control of the developers. So although a process may appear fixed and unyielding to variation, in practice it must adopt some "agile" approaches.

The first formal description of the waterfall model was in an article by Winston W. Royce [Royce 1970]. Although there has been much criticism of a "pure" waterfall model, almost all software development models bear some similarity to the waterfall model. As we shall see in the other models discussed in this mini-tutorial, they all incorporate at least some phases similar to those used in the waterfall model.

Spiral Model

The spiral model, developed by Barry Boehm [Boehm 1988] and as depicted Figure 2.2, is a software development process, which combines the features

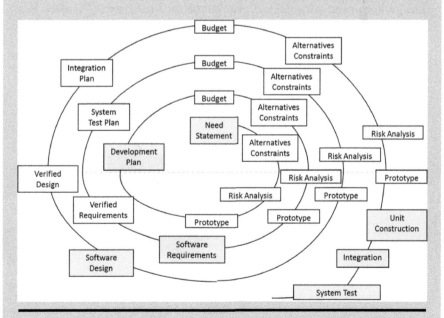

Figure 2.2   Spiral model.

of the prototyping and the waterfall model. Starting at the center, each turn around the spiral goes through several task regions:

- Determine the objectives, alternatives, and constraints for each new iteration.
- Evaluate alternatives and identify and resolve risk issues.
- Develop and verify a prototype and product artifact for the iteration.
- Plan the next iteration.

Although the waterfall features of the Spiral Model might prompt a plan-driven approach, the cyclic nature of the Spiral, along with the evolution of prototypes, represents agile-type features.

V-Model

The V-Model, depicted in Figure 2.3 is a SDLC model that represents a restructuring of the waterfall model, with the non-testing software phases of the life cycle on the left side of the V and the testing phases on the right side of the V. The V-Model uses the double arrowhead dashed line to show the relationship between developing a software artifact (Need Statement, Requirements Specification, an Architectural Design document, Module Design document) with the test that will be used to verify the artifact. For

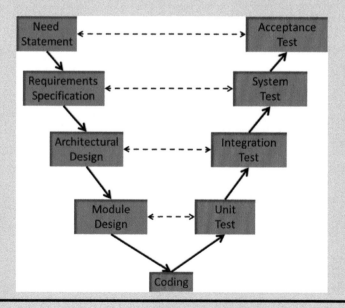

Figure 2.3   V-Model.

example, when a software architecture for a system is determined, developers can, and they should, start planning for the test of integration of the architectural components. Likewise, when conducting an integration test, the defects discovered typically point to problems in the architecture. So, this is a two-way street.

Chapter 3 discusses the various testing phases, including their purpose, planning, execution, and analysis.

Incremental Model

Figure 2.4 illustrates the Incremental Model. The figure shows that a software system is built in increments and assembled as each increment is completed. There are a number of advantages of this approach:

- It uses a "divide and conquer" to reduce scope and complexity, allowing developers to concentrate on just a subset of the problem to be solved.
- Since the scope of the first increment is necessarily less complex than the whole development activity, it provides an opportunity for project team to do some experimentation (e.g., use of a new tool or methodology) and learn new approaches, without great risk to the entire project.
- The first increment also provides the opportunity for the project team to establish effective working relationships and communication mechanisms, without being overwhelmed by the size and complexity of the entire system development.

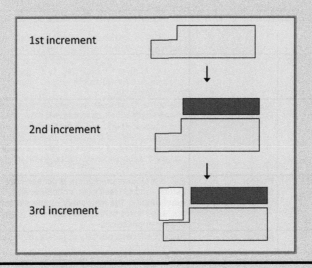

Figure 2.4   Incremental model.

- Depending on the nature of the increments, customers may be able to observe working portions of the system, providing timely feedback to developers and giving the customers a clear view of project progress.

An incremental approach is widely used by software developers and is incorporated in many software processes.

Capability Maturity Model

In the late 1980s, the Software Engineering Institute (SEI), at Carnegie Mellon University, developed the SW-CMM (Software Capability Maturity Model) to help organizations build effective software engineering processes [Paulk et al. 1993]. The SW-CMM and its successor the CMMI (Capability Maturity Model Integration) have been widely adopted in industry, primarily by large and moderate sized software development organizations.

The CMM is not a software process model, like the above models, but is rather a software process framework or a reference model of mature practices, which is organized into a five-level model encompassing good engineering and management practices. Figure 2.5 describes the five levels of maturity practices.

This framework is intended to help software organizations improve the maturity of their software processes in terms of an evolutionary path from ad hoc, chaotic processes to mature, disciplined software processes. Although some aspects of the model may be involved in agile processes [Paulk 2001], it has primarily been used in organizations that adopt a plan-driven approach.

Level	Focus	Decription
5: Optimizing	Continuous Process Improvement	Continuous process improvement is enabled by quantitative feedback from the process and from piloting innovative ideas and technologies.
4: Quantitatively Managed	Product and Process Quality	Detailed measures of the software process and product quality are collected. Both the software process and products are quantitatively understood and controlled.
3: Defined	Engineering Process	The software process for both management and engineering activities is documented, standardized, and integrated into a standard software process for the organization. All projects use an approved, tailored version of the organization's standard software process for developing and maintaining software.
2: Managed	Project Management	Basic project management processes are established to track cost, schedule, and functionality. The necessary process discipline is in place to repeat earlier successes on projects with similar applications.
1: Initial	No Focus	Project success primary depends on individuals and their heroics.

Figure 2.5   SW CMM maturity levels.

SPECIFIC SOFTWARE PROCESSES

In this section, we describe specific defined software processes that are used by project teams in the life cycle development of software. All the processes are customizable and are typically adapted to the project application domain and to an organization's structure and culture. Although our discussion is not exhaustive, it illustrates a range of plan-driven and agile methods.

Unified Process

The Unified Process (UP) is an iterative and incremental software development process model [Scott 2002]. The best-known refinement of the UP is the Rational Unified Process [Kruchten 2004]. Figure 2.6 displays the UP phases:

- *Inception* – Define the scope of the project and develop the business case.
- *Elaboration* – Plan project, specify features, and baseline the architecture.
- *Construction* – Build the product.
- *Transition* – Transition the product to its users.

UP practices include the following:

- Iterative development within the Inception, Elaboration, Construction, and Transition phases
- Each iteration through each of the phases results in an "increment" release
- Use cases, discussed in Chapter 5, are used to capture the functional requirements and to define the contents of the iterations
- Architecture provides the focus for system development efforts to the most critical risks that are addressed in the early phases of development

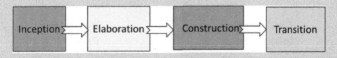

Figure 2.6 UP process.

eXtreme Programming Process

The eXtreme Programming (XP) process is an agile process designed for work in small to medium teams (less than ten members), building software with vague or rapidly changing requirements, and it advocates frequent "releases" in short development cycles, which is intended to improve productivity and introduce checkpoints where new customer requirements can be adopted. The XP life cycle has four basic activities [Beck 2000]:

- Continual communication with the customer and within the team
- Simplicity, achieved by a constant focus on minimalist solutions
- Rapid feedback through unit and functional testing
- Emphasis on dealing with problems proactively

Figure 2.7 depicts XP practices, some of which are described as follows:

- *Planning Game* – Quickly determine the scope of the each release by combining business priorities and technical estimates. The customer decides scope, priority, and dates from a business perspective, whereas technical people estimate and track progress.
- *Small Releases* – Puts a simple system into production quickly. Release new versions on a very short (2 week) cycle.

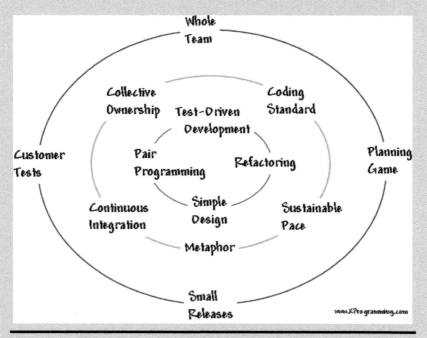

Figure 2.7   XP practices.

- *Metaphor* – Guide all development with a simple shared story of how the overall system works.
- *Test-Driven Development* – Developers continually write unit tests (before unit design and code) that must run flawlessly for development to continue; customers write tests to demonstrate that features are finished.
- *Refactoring* – Programmers restructure the system without changing its behavior to remove duplication, improve communication, simplify, or add flexibility.
- *Pair Programming* – All production code is written by two programmers at one machine.
- *Continuous Integration* – Integrate and build the system many times a day, every time a task is completed.
- *Collective Ownership* – Anyone can change any code anywhere in the system at any time.
- *Coding Standards* – Programmers write code in accordance with rules that emphasize communication throughout the code.

XP introduced a number of new practices and terms (pair programming and refactoring), but also uses certain plan-driven aspects (Planning Game and Coding Standards) and builds on the incremental model (Small Releases and Continuous Integration).

SCRUM PROCESS

The Scrum process is an empirical process – accepting that the problem cannot be fully understood or defined, focusing instead on maximizing the team's ability to deliver quickly and respond to emerging requirements. Scrum concentrates on project management without dictating other software development practices (such as analysis, design, construction, and testing) [Schwaber 2004]. Figure 2.8 outlines the key features of the Scrum.

A *sprint* is the basic unit of development in Scrum. The sprint duration is fixed in advance for each sprint and is normally between one week and one month. Each sprint is preceded by a *planning meeting*, where the tasks for the sprint are identified and an estimated commitment for the sprint goal is made. This is followed by a *review* where the team and the customer review and evaluate the product, and finally a *retrospective* meeting, where the process is assessed is reviewed and lessons and continuous improvement opportunities for the next sprint are identified.

The set of features that go into a sprint come from the *Product Backlog*, which is an ordered list of requirements. Which backlog items go into the sprint (the sprint goals) is determined during the sprint planning meeting.

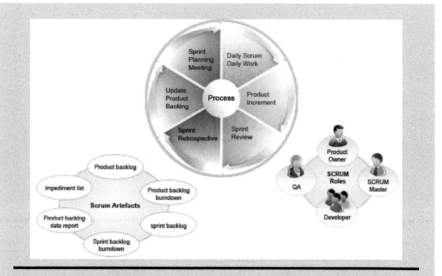

Figure 2.8   Scrum features.

Each workday, the Scrum team holds a *Daily Scrum*: the meeting lasts for 15 minutes and is held in the same location and at the same time; all members give updates; all are welcome, but normally only the core roles speak. The *Scrum Master* asks each team member three questions, in turn:

- "What have you done since the last meeting?"
- "What do you plan to do before the next meeting?"
- "What impediments are in your way?"

Team Software Process

The Team Software Process (TSP) was developed by the SEI and includes most of the key process areas of the SW-CMM [Humphrey 2000, Davis 2003] and is considered a Level 5 process, at least at the team level (see Figure 2.4). The objective of the TSP is to create a team environment that supports disciplined individual work and builds and maintains a self-directed team.

Software is developed incrementally in multiple cycles, with each cycle made up of multiple phases. Figure 2.9 illustrates n-cycle version. Each cycle represents a life cycle development of some component of the final product with previous components integrated in each cycle.

The TSP focuses on team-building, team-management, team member capability, quality assurance, and software process improvement. The defined team roles promote shared team management and a collegial work environment (e.g., Planning Manager and Support Manager). Although TSP uses mostly plan-driven methods, it does exhibit some agile features [Paulk 2001].

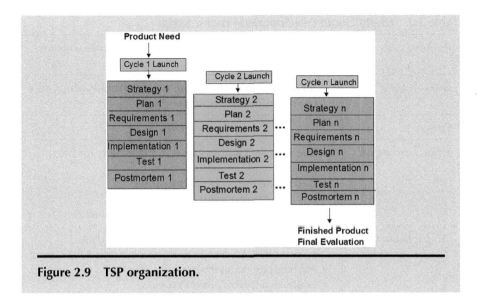

Figure 2.9   TSP organization.

At the completion of the Software Process Fundamentals MiniTutorial, Massood approached Jose during one of the breaks and discussed the possibility of using an agile process (such as Scrum) for the development of the prototype. Jose had recently heard about the Scrum process and its adaptation in mainly software industries but was not really familiar with the process and its advantages and disadvantages. Since Massood was a new hire and the youngest member of the team, Jose did not want to ignore his suggestion. He was also intrigued by the idea that the most junior member of the team had taken initiative and proposed an alternate approach. Therefore, Jose told Massood that he would like him to prepare and deliver several Scrum MiniTutorials to educate the team about Scrum. This made Massood very happy, because he realized *HomeOwner* valued the opinion of every member of the team.

NOTE: The collection of the Scrum process MiniTutorials can be found in Chapter 10.

EXERCISE 2.2: EVALUATING A SOFTWARE PROCESS

The Project Launch Process (Table 2.1) specifies the activity "Establish the DH Development Process". So part of the two-day launch workshop was devoted to this topic.

Prior to the workshop, Jose Ortiz had prepared a draft of the *DigitalHome Development Process Script* (Table 2.2) to guide the DH project work. He requested that the DH project team review the process script and provide him with an evaluation of its viability, with comments and recommendations.

The exercise, described at the end of this chapter, deals with the evaluating the draft DH Development Process.

Table 2.2   DigitalHome Development Process Script

Purpose	• Guide the development of the *DigitalHome* software system.
Entry Criteria	• Identification of a problem or opportunity, which may need a software solution.
Phase	*Activity*
Project Inception	• Establish the business case and feasibility for the project • Determine the scope of the project • Develop a customer need statement • Develop a high-level requirements document
Project Launch	• Establish development team, and their roles, responsibilities, and goals. • Analyze Need Statement • Establish a Development Process • Identify External Interfaces and create Context Diagram • Create a Conceptual Design • Determine a Development Strategy • Research and study of possible project technology
Planning (discussed in Chapter 4)	• Develop Task, Schedule, and Resource Plans • Develop a Risk Plan • Develop a Quality Plan • Develop a Configuration Management Plan • Determine project standards and tools • Perform project planning/tracking
Analysis (discussed in Chapter 5)	• Review the Need Statement and prepare for requirements elicitation • Elicit requirements from potential customers and users • Analyze requirements and build object-oriented analysis models • Specify and document requirements in an SRS • Inspect and revise the SRS • Obtain customer agreement to requirements • Develop a System Test Plan • Develop Requirements Traceability Matrix (RTM) • Develop User Manual • Track, monitor, and update plans (Task/Schedule/ Quality/CM/Risk) & RTM

(Continued)

Table 2.2 (*Continued*)   DigitalHome Development Process Script

Purpose	• Guide the development of the *DigitalHome* software system.
Entry Criteria	• Identification of a problem or opportunity, which may need a software solution.

Phase	*Activity*

Architectural Design (discussed in Chapter 6)	• Develop OO high-level design models
	  • Refine the OO analysis models
	  • Develop additional design models
	  • Specify and document design in an SDS (Software Design Specification)
	• Develop a Build/Integration Plan
	• Review and revise the SDS
	• Consult with customer, and revise SRS and SDS, as appropriate
	• Track, monitor, and update plans (Task/Schedule/ Quality/CM/Risk) & RTM

Construction Increment 1, 2,... (discussed in Chapter 7)	• Complete Class Specifications (data types and operation logic) • Develop Increment (Class/Integration) Test Plan • Set up appropriate test environment (e. g, test harness, stubs/drivers) • Review the Class Specifications • Write source code • Review and test the source code • Integrate code with previous increments • Perform Increment (Class/Integration) Test • Consult with customer, and revise SRS, SDS, and code, as appropriate • Track, monitor and update plans (Task/Schedule/ Quality/CM/Risk) & RTM
System Test (discussed in Chapter 3)	• Update test environment • Update System Test Plan • Perform the System Test • Consult with customer, and revise requirements, design, and code as appropriate • Track, monitor, and update plans (Task/Schedule/ Quality/CM/Risk) & RTM

(*Continued*)

Table 2.2 (*Continued*) DigitalHome Development Process Script

Purpose	• Guide the development of the *DigitalHome* software system.
Entry Criteria	• Identification of a problem or opportunity, which may need a software solution.
Phase	*Activity*
Acceptance Test (discussed in Chapter 3)	• Prepare Acceptance Test materials (including a user's manual) • Set up customer test environment • Customer performs Acceptance Test • Revise SRS, SDS, code, and User Manual as appropriate • Track, monitor, and update plans (Task/Schedule/ Quality/CM/Risk) & RTM
Postmortem	• Perform postmortem analysis of project process and product quality
Exit Criteria	• Completed SRS, SDS, RTM, User Manual, System Test Plan • Source code for a thoroughly reviewed and tested program • Completed plan documentation • Project postmortem report

Development Strategy

On the second day of the DH Launch Workshop, Michel Jackson, the DH System Analyst, led the discussion of the DH Conceptual Design. Michel explained that they would be developing a Context Diagram for the DH system and a Conceptual Design, presenting a high-level view of the principle components for the DH system.

Massood Zewail said he understood the role of the context diagram (a view of how the system as a whole interacts with external entities) and why it was developed early in the process, but he did not understand why we would be doing design work prior to specifying the DH requirements. He was taught in his master's program that developers should resist thinking about the design before the requirements are established – "don't try to solve the problem before you understand it".

Michel agreed that they should not try to design a solution before the DH requirements had been analyzed and specified. However, the "Conceptual" Design would not be an actual solution, but a rather a pre-planning document that would make it easier for the team to establish a development strategy and prepare a detailed project plan. The Conceptual Design would be like a "throwaway" user interface prototype: a tool used to show a user what the system looks like in order to elicit requirements, but thrown away once the requirements are discovered. Another advantage of the conceptual design is the fact that it will be used for project size and effort estimation.

Georgia Magee chimed in with a comment about her work at the Volcanic Power Company, where she was a member of an XP team. Georgia described the "Planning Game" method they used, early on the project, to develop a strategy for planning the project iterations; she thought was very helpful and agreed with Michel's comments. Yao Wang added that he had worked on team that used the UP process, and in the Inception phase they developed a "candidate architecture", which had seemed similar to the Conceptual Design idea.

Massood said "okay, I see the rationale for a conceptual design - thanks".

In the next two hours, the DH Team developed the Context Diagram in Figure 2.10 and the Conceptual Diagram in Figure 2.11. After this, Disha Chandra led the team in generating a DH Development Strategy. Disha explained that in the next phase of the project, the Planning Phase, they would develop a detailed list of tasks, estimate the amount of effort required of each task, scheduled the tasks, and assign lead responsibility. She also pointed out that their development process (Table 2.2) includes multiple construction increments, which raises questions about what will be developed in each increment and how long each increment will take to complete.

Disha stated that the team needs a strategy for development to do project planning. That is, a strategy that sets down how many increments will be developed, what will be constructed in each increment, and provides estimates of the size and effort required for each increment.

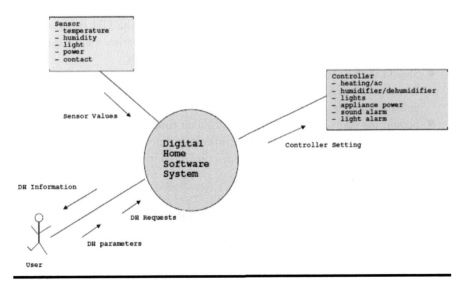

Figure 2.10 DH context diagram.

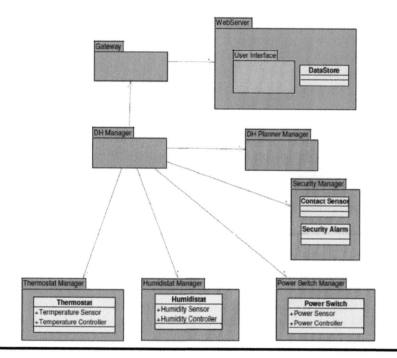

Figure 2.11   DH conceptual design.

So, how do we start? Disha asserted that the central issue is what criteria will be used to determine the Development Strategy:

- Should the first increment be a minimum subset of the DH functionality? This could provide the opportunity for the team to experience some early success, avoiding the risk of failure by taking on too much.
- Should the first increment provide a base for DH construction that can easily be enhanced in future increments, making integration of components easier, perhaps a walking skeleton?
- Should the team start with the most difficult functionality, solving the most difficult problems in the beginning, allowing time during subsequent increments to correct errors and to enhance the initial solution?
- Or something else?

EXERCISE 2.3: DH DEVELOPMENT STRATEGY

The team then proceeded to produce a strategy criterion and an allocation of conceptual design components to the various construction increments.

The exercise, described at the end of this chapter, deals with generating a DH Development Strategy.

Case Study Exercises

EXERCISE 2.1: FORMING AN EFFECTIVE TEAM

Jose Ortiz and Disha Chandra felt that after the team building tutorial the team needed to start working on forming a cohesive and effective team. The earlier Team Problems activity (see Chapter 1) was a good start, but they both felt an additional Team Formation activity was needed.

LEARNING OBJECTIVES

Upon completion of this exercise, students will have increased ability to:

- Discuss some of key issues in building effective teams
- Develop techniques for building effective teams
- Appreciate the need for project team members to work together effectively

EXERCISE DESCRIPTION

1. As preparation for the exercise, read Chapter 1.
2. You will be assigned to a small group.
3. Each group designs a team building activity to help the team to coalesce and become a cohesive, effective team. The team building activity should follow the below guidelines:
   - Can be completed in 45 minutes by a team of 4 or 5 people. Possibly followed by 15-minute class discussion.
   - Ensures all team members participate.
   - Concerns something related to the team's work on its project and produces something useful to the team.
4. Each team summarizes its discussion and conclusions in a written report. One member of their group is chosen to report to the class on the group's discussion and the team building activity it designed.

EXERCISE 2.2: EVALUATING A SOFTWARE PROCESS

Jose Ortiz has asked the project team to review the draft of the *DigitalHome Development Process Script* (Table 2.2). He requested that they provide him with an evaluation of its viability, with comments and recommendations.

LEARNING OBJECTIVES

Upon completion of this exercise, students will have increased ability to:

- Understand and analyze the purpose and implementation of software processes.
- Evaluate the fundamental elements of the content and structure of a software process for a given project.

EXERCISE DESCRIPTION

1. As preparation for the exercise, read Chapter 1.
2. You will be assigned to a small group.
3. Each team evaluates the DH Development Process Script. The following questions should be considered in the evaluation:
   - Is this an agile or a plan-driven process?
   - What are the strengths and weaknesses in the process?
   - Is the process appropriate for the domain area, development context, and delivery schedule?
   - Is the process clear and easy to understand? Does it have sufficient detail? Or is it too detailed?
   - Is the process complete? Are there any missing phases or activities?
   - Are there ways that the process could be improved with changes or additions or deletions?
   - Which phases should be assigned to individual DH Team members to lead?
4. Each team summarizes its discussion and conclusions in a team evaluation report. Choose one member of each group to report to the class on the team's evaluation.

EXERCISE 2.3: CREATING A DH DEVELOPMENT STRATEGY

During the Launch Workshop Disha Chandra led the DH Team in creating a DH Development Strategy. This exercise concerns this activity.

LEARNING OBJECTIVES

Upon completion of this exercise, students will have increased ability to:

- Discuss the purpose and need for a Development Strategy.
- Develop a Development Strategy for a software project.
- Assess the adequacy of a Conceptual Design to support creating a Development Strategy

EXERCISE DESCRIPTION

1. As preparation for the exercise, study the DH Context Diagram and Conceptual Design, and the section of this Chapter on the Development Strategy.
2. You will be assigned to a small team.
3. Each group carries out the following activities:
   - Determine the Development Strategy Criteria.
   - Determine the number of development increments.
   - Allocate conceptual design modules to each increment.
   - Estimate module size and development effort for each increment.
4. Each team documents their Development Strategy for DH in the below template. One member of their group is chosen to report to the class on the group's discussion and resulting strategy

DigitalHome Development Strategy			
Team Names			
Strategy Criteria			
Increment #	*Module/Element*	*Size (LOC)*	*Effort (days)*
1	…		
1	…		
…			
2	…		
2	…		
…			

Chapter 3

Assuring *DigitalHome* Quality

Software Quality Assurance

After the project team had decided on their development process (Table 3.1), Jose Ortiz asked the team to remember the importance that *HomeOwner* management had placed on the need for a high-quality *DigitalHome* prototype. Massood Zewail, who is a recent graduate of a Master of software engineering program, affirmed the importance of software quality assurance (SQA) and suggested that the team study the V-Model software development process for its features, which would help ensure a high-quality product. Some members of the team were not familiar with the V-Model, and some were not convinced its quality features could be properly integrated into the Table 3.1 process. Also, Yao Wang expressed concern about ensuring the security of *DigitalHome* system. Jose asked Massood to prepare a tutorial on SQA and include discussion of the V-Model security engineering. Massood agreed to deliver a tutorial on quality assurance; however, he also pointed out that "quality without measurement is just cheerleading"; therefore, he also included some discussion of measurement as part of his presentation.

Table 3.1 *DigitalHome* Development Process Script

Purpose	• Guide the development of the *DigitalHome* software system
Entry Criteria	• Identification of a problem or opportunity, which may need a software solution
Phase	*Activity*
Project Inception	• Establish the business case and feasibility for the project • Determine the scope of the project • Develop a customer need statement • Develop a high-level requirements document
Project Launch	• Establish development team, and their roles, responsibilities, and goals. • Analyze Need Statement • Establish a Development Process • Identify External Interfaces and create Context Diagram • Create a Conceptual Design • Determine a Development Strategy • Research and study of possible project technology
Planning (discussed in Chapter 4)	• Develop Task, Schedule, and Resource Plans • Develop a Risk Plan • Develop a Quality Plan • Develop a Configuration Management Plan • Determine project standards and tools • Perform project planning/tracking
Analysis (discussed in Chapter 5)	• Review the Need Statement and prepare for requirements elicitation • Elicit requirements from potential customers and users • Analyze requirements and build object-oriented analysis models • Specify and documents requirements in an SRS • Inspect and revise the SRS • Obtain customer agreement to requirements • Develop a System Test Plan • Develop Requirements Traceability Matrix (RTM) • Develop User Manual • Track, monitor and update plans (Task/Schedule/Quality/CM/Risk) & RTM

(Continued)

Table 3.1 (*Continued*)   *DigitalHome* Development Process Script

Purpose	• Guide the development of the *DigitalHome* software system
Entry Criteria	• Identification of a problem or opportunity, which may need a software solution

Phase	*Activity*
Architectural Design (discussed in Chapter 6)	• Develop OO high-level design models • Refine the OO analysis models • Develop additional design models • Specify and document design in an SDS (Software Design Specification) • Develop a Build/Integration Plan • Review and revise the SDS • Consult with customer, and revise SRS and SDS, as appropriate • Track, monitor and update plans (Task/Schedule/Quality/ CM/Risk) & RTM
Construction Increment 1, 2,... (discussed in Chapter 7)	• Complete Class Specifications (data types and operation logic) • Develop Increment (Class/Integration) Test Plan • Set up appropriate test environment (e.g., test harness, stubs/drivers) • Review the Class Specifications • Write source code • Review and test the source code • Integrate code with previous increments • Perform Increment (Class/Integration) Test • Consult with customer, and revise SRS, SDS, and code, as appropriate • Track, monitor and update plans (Task/Schedule/Quality/ CM/Risk) & RTM
System Test	• Update test environment • Update System Test Plan • Perform the System Test • Consult with customer, and revise requirements, design and code as appropriate • Track, monitor and update plans (Task/Schedule/Quality/ CM/Risk) & RTM

(Continued)

Table 3.1 (*Continued*)   *DigitalHome* Development Process Script

Purpose	• Guide the development of the *DigitalHome* software system
Entry Criteria	• Identification of a problem or opportunity, which may need a software solution
Phase	*Activity*
Acceptance Test	• Prepare Acceptance Test materials (including a user's manual) • Set up customer test environment • Customer performs Acceptance Test • Revise SRS, SDS, code, and User Manual as appropriate • Track, monitor, and update plans (Task/Schedule/Quality/CM/Risk) & RTM
Postmortem	• Perform postmortem analysis of project process and product quality
Exit Criteria	• Completed SRS, SDS, RTM, User Manual, System Test Plan • Source code for a thoroughly reviewed and tested program • Completed plan documentation • Project postmortem report

MINITUTORIAL 3.1: QUALITY ASSURANCE THROUGHOUT SOFTWARE DEVELOPMENT LIFE CYCLE

WHY QUALITY?

Quality assurance should be an integral part of software development and be embedded in every phase of the development life cycle. By controlling quality during software development, there is a better chance of meeting the project schedule and budget. In addition, software quality has become a competitive issue and is essential for survival of a development organization. By producing high-quality software products, an organization can retain its existing customers, and can potentially gain additional customers from its competitors.

Achieving high quality requires buy in from every member of the organization. Managers must provide appropriate training and tools to enable the employees to achieve the quality goals. Also, managers need to establish project plans that provide sufficient time for the project team to develop a quality product. Unfortunately, some managers view activities associated with quality assurance as non-productive activities; however, industry data has shown that quality assurance activities, throughout the development life cycle, result in significant savings.

SQA is "a set of activities that define and assess the adequacy of software processes to provide evidence that establishes confidence that the software processes are appropriate for and produce software products of suitable quality for their intended purposes" [IEEE 730]. *Software quality* is typically expressed in one of two ways: *defect rate* (e.g., defects per lines of source codes, per function point, or per some other unit) and *reliability* (number of failures per some unit of time, mean time to failure, or probability of failure-free operation in some specified time) [Kan 2002].

Numerous studies have shown that SQA has played a major role in the overall cost and schedule of software projects [Davis 2003, Dehaghani 2013, Parnas 2003, Standish 2020]. In the early years of the software development industry, SQA was synonymous with software testing. As a result, since the waterfall model visually presented the development process as a series of sequential phases (see Figure 3.1), this promoted the idea that software quality activities would be conducted after most of development was complete.

However, industry data has revealed that 50%–60% of project defects are introduced during the requirements phase [Marandi 2014]. Of course, there are problems with not detecting and removing defects as early as possible. First, every defect that is not found at the phase in which it is introduced has the potential to introduce up to ten additional defects in the following phases of development. Therefore, a defect introduced during the requirements phase has

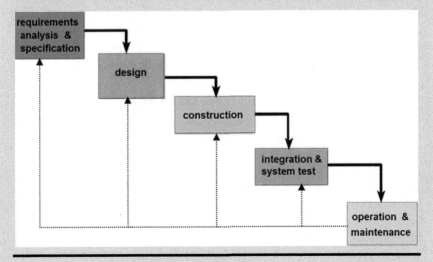

Figure 3.1   Waterfall process model.

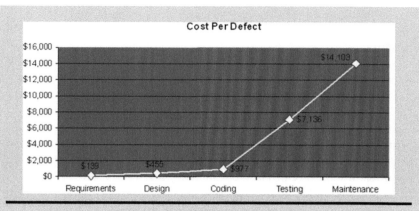

Figure 3.2   Defect removal cost increase.

the potential to introduce numerous defects during the implementation phase. The second problem with leaving defects in the product at the earlier phases of the development is that the cost associated with its removal in later phases is much higher than the cost that would be required to remove it in the hase in which it was introduced (see Figure 3.2) [Jones 2003]. For example, imagine a defect that is introduced during the requirements phase, which might easily be detected (we discuss different detection techniques later in this chapter) and corrected with a word processor. If that defect is detected during the testing phase, it might require a change in design and implementation at a much higher cost.

Over the years, as the field of software engineering has matured, alternative development life cycles have been introduced, with each providing additional advantages and disadvantages over the waterfall model. Different development life cycles are discussed in Chapter 2; however, we will discuss a development life cycle model called the "V-Model", in which quality is a prominent.

THE "V-MODEL" DEVELOPMENT LIFE CYCLE

The V-Model development life cycle model, which was discussed briefly in Chapter 2, emphasizes software quality activities. As represented in Figure 3.3, the V-Model contains seven square boxes, where each box represents some activity associated with the development and/or quality assurance of the software product. There are two bubbles, where each bubble represents a quality review activity that is performed on the artifacts that are generated from the activities performed at each of the square boxes. There are three paths that are identified as part of this life cycle: the development phase (yellow boxes and solid blue arrows), the quality assurance planning path (the orange dash-dot lines connecting the green boxes), and the quality assurance execution path (the green boxes and the light blue dash-dot lines). The blue dot line shows

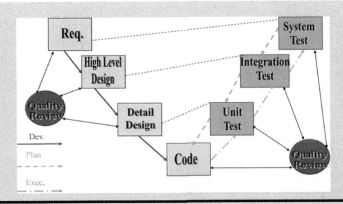

Figure 3.3   V-Model development life cycle.

the bi-directional relationship between the yellow and green boxes. Finally, at each phase, there are quality review activities (inspection, walk-through).

The key idea behind the V-Model is the fact that quality assurance has been integrated throughout the development life cycle. There is a quality review that is performed on each artifact that is generated as part of the development life cycle. In addition, as the requirements are being defined, the system test plan is also created. The same is true about the high-level design and detail design, which are in parallel with development of the integration and unit test plans, respectively. Of course, as source code is created the unit tests are executed, and as design modules are integrated, the integration test is executed, and finally the system test is executed at the completion of the product or sub-product. This process supports Test-Driven Development, which promotes the concept of designing the test, before writing any code.

SECURITY ENGINEERING

An essential element of quality assurance is to protect the security of a system [Pressman 2015]. With the development of a system such as *DigitalHome*, developers must be aware of potential attacks on user privacy and *DigitalHome* information. Security engineering consists of methods and practices that help protect from such attacks and ensure the security of a system.

The following are key techniques used to protect the security of a system:

■ *Security Risk Management* – identify, analyze, and monitor security risks. Chapter 4 contains a section on risk management which covers a spectrum of risks, including security risks.
■ *Security Requirements Engineering* – elicit, analyze, specify, and validate security requirements. Chapter 5 discusses requirements activities for a range of requirements types, including security requirements.

■ *Security Design* – develop an architecture and design modules, which help ensure security requirements are properly implemented. Chapter 6 discusses design concepts, such as Modularity, Abstraction, Information Hiding, High Cohesion, and Low Coupling, which support development of a system which addresses security issues.

■ *Secure Coding* – use of coding practices such as adopting a secure coding standard, validating input, rigorous access denial, and implementing multiple defensive strategies. Chapter 7 covers this area.

■ *Security Verification* – use of verification techniques such as requirements, design and code reviews, and software testing to check for security problems. These techniques are covered later in this chapter.

EXERCISE 3.1: SECURITY ATTACKS

After the discussion of the "Security Engineering", Massood led the team in a discussion about potential attacks on the security of the *DigitalHome* system.

The exercise, described at the end of this chapter, deals with the DH Team's discussion.

THE DEFECT LIFE CYCLE

■ As shown in Figure 3.4, developers introduce, through a mistake or an error, a defect (or fault) into a product. If the fault is not exposed during program execution, there would be no external evidence of its existence; however, as soon as that defect/fault is executed, it results in a failure. For example, imagine a mistake of using a "<" rather than ">" in the control of thermostat setting could produce a problem with turning on the air conditioning system during a hot summer day. Understanding this "defect life cycle" is critical in the overall perceived quality of the product. A product may have many defects embedded in it, and still be perceived as a quality product, as long as none of those defects are exposed in execution. Therefore, it is very important to establish appropriate quality control strategies to detect and remove defects that have the potential to be executed during operation. This may mean prioritizing the defects and removing the defects that have the highest probability of execution. This is an acceptable strategy and practice, as it is almost impossible to remove all the defects from a large, complex system.

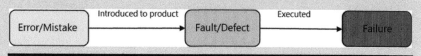

Figure 3.4   Defect life cycle.

COST OF QUALITY

There are several factors that influence the overall cost of software quality:

- the overall complexity and size of the product;
- the criticality of the product (i.e., as the cost of failure increases, the quality cost increases);
- the inherent risk of the product (i.e., typically there is a higher quality cost associated with high risk projects, such as a real-time embedded systems);
- the competency of the developers and quality assurance personnel;
- product stability/volatility (i.e., the more volatile product typically results in a lower quality product); and
- the availability and the efficiency of the development and testing infrastructure (i.e., automation tools can reduce the overall cost of quality).

The overall maturity of the organization and its personnel will have a direct relationship with the overall quality costs of the product.

There are number of factors that contribute to the cost of quality activities:

- *Prevention Cost* – this includes all the costs, which are associated with activities by the project team to prevent introduction of defects into the product. Establishing standards, processes, and infrastructure are contributors to this cost.
- *Appraisal Cost* – this includes all the costs, which are associated with activities by project team to appraise the quality of the product. These are activities that are associated with the detection of the defects in the product. Inspections and testing are examples of such activity.
- *Failure Cost* – this includes the cost that is associated with the activities, which result from a product failure. There are two different categories of failure costs: internal and external. An example of an internal failure cost is the rework cost associated with the failure that is detected during software testing. Examples of external failure costs include the maintenance cost after the product release, the recall associated with the faulty product, or the litigation cost that is associated with the failed product.

Given the competitive nature of the software industry, and the effect of the software products on day-to-day activities, it is necessary to invest an appropriate level of resources throughout the development life cycle to achieve the desired level of quality. The key phrases in this statement are the "appropriate level of resources" and "desired level of quality". Unfortunately, in most cases, there are not enough resources available in a project to build an error-free product, and it is rare that a complex product is free of error. One of the biggest challenges in quality assurance is to find the right balance between the desired level of quality, and appropriate level of resources. The next section describes activities that can help reach the right balance.

EXERCISE 3.2: COST OF QUALITY EXPERIENCES

After the discussion of the "Cost of Quality", Massood lead the team in a discussion about their experiences with the types of costs of quality they had experienced in their previous software development work.

The exercise, described at the end of this chapter, deals with the DH Team's discussion.

Software Quality Assurance Processes

Massood Zewail also prepared and delivered tutorial on the processes used in SQA.

MINITUTORIAL 3.2: SOFTWARE QUALITY PROCESSES

There are five major processes associated with SQA; these are planning, oversight, record keeping, analysis, and reporting [IEEE 730].

- During the **Planning** activities, the organization identifies a set of quality goals and develops a roadmap on how to achieve those quality goals. The roadmap includes the purpose and the scope of the quality activities, and the required resources and processes that need to be implemented to achieve those quality goals. In addition, roles and responsibilities of the people involved in quality assurance activities are defined. In some organizations, especially for large complex projects, SQA personnel are separate from the development team. A SQA plan is an artifact that is generated as the result of planning activities. Such a plan would be the type of plan specified as a "Quality Plan" in Planning section of the *DigitalHome* Development Process Script in Table 3.1.
- During the **Oversight** activities, the quality assurance personnel (which may or may not include the development team) will monitor the software development process, and the resulting product to make sure all the processes are followed, and the quality goals are met. One major component of the oversight is measurement and data collection.
- **Record Keeping** requires identification of the data that is going to be collected, the tools and procedures that will be used to facilitate data collection and storage, and the roles and responsibilities associated with the record keeping.

- Once the data is collected and stored, the **Analysis** process begins. Through the data analysis the quality assurance personnel and the developers will be able to monitor the success and failure of the quality assurance processes. The analysis will reveal the opportunities for continuous improvement.
- The information collected as part of the quality assurance activities is valuable data that needs to be shared with different constituents. **Reporting** of the data must include what data is to be collected and who will receive it. Who will receive data needs to be identified early in the quality assurance process and communicated with every member of the team (developers and quality assurance people). Such communication is part of confidence building between the quality assurance people, developers, and managers.

SOFTWARE QUALITY ASSURANCE PLAN (SQAP)

The purpose of the SQAP is to establish the scope, goals, processes, resources, and responsibilities that are required in to implement an effective quality assurance function. The SQAP is a quality plan that incorporates the elements of SQA activities discussed in the previous section in Table 3.1. The *IEEE Standard for Software Quality Assurance Process* [IEEE 730] specifies that a SQAP document should contain the following sections:

- *Purpose* – identifies the specific purpose and scope of the SQA activities, including the portion of the software and the specific phases of the development life cycle that are within the scope of these SQA activities.
- *References* – lists documents that are needed throughout for the SQA activities.
- *Management* – describes the project's organization structure, its tasks, and its roles and responsibilities.

■ *Documentation* – identifies the set of documents (i.e., requirement, design, etc.) that fall within the scope of the SQA activities. This section will lay out a plan on how each document will be assessed and approved. In addition, this section will identify the set of documents that will be generated as part of the SQA activities.

■ *Standards and Metrics* – describes the set of standards and metrics that will be used throughout the quality assurance activities.

■ *Reviews and Audits* – defines the different reviews and audits that will be conducted throughout the quality assurance activities; this includes what reviews and audits will be conducted, how they will be accomplished, and how the results of those audits and reviews will be used.

■ *Testing* – describes the overall activities that are associated with the software testing. This includes, making sure that the tests are conducted on time, follow the defined process, and appropriate data is collected.

■ *Problem Reporting and Corrective Actions* – specifies how problem reports are collected, who will see those data, and what procedures should be followed for implementing the appropriate corrective actions.

■ *Training* – identifies the necessary training, if any, that is needed to make sure that the people involved in the quality assurance process are able to conduct their roles and responsibilities effectively.

EXERCISE 3.3: SOFTWARE QUALITY ASSURANCE PLANNING

After Massood's discussion of SQA planning, Disha Chandra lead the team in a discussion SQA planning.

The exercise, described at the end of this chapter, deals with the DH Team's discussion.

Quality Measurement and Defect Tracking

Jose Ortiz asked Massood Zewail if he would also prepare and deliver a tutorial on the processes used in software quality measurement and tracking.

MINITUTORIAL 3.3: QUALITY MEASUREMENT AND DEFECT TRACKING

Evidence of the presence, absence, or variance from the expected quality of a software assurance activity can be determined through measurement. Establishing a solid measurement strategy is one of the most important components of any quality assurance organization. As the *IEEE Standard for a Software Quality Metrics Methodology* [IEEE 1061] states "...defining software quality for a system is equivalent to defining a list of software quality attributes required for that system. In order to measure the software quality attributes, an appropriate set of software metrics shall be identified". When planning for a measurement strategy, the following questions are appropriate:

- Why do we need to measure?
- What do we need to measure?
- How do we conduct these measurements?
- Who will be looking at these data?
- What do we do with the data collected?

The answer to these questions provides guidelines for the establishment of the metrics that are needed, the personnel that are going to conduct these measurements, the infrastructure that are needed to collect and maintain the data, the necessary analysis of the data, and personnel who will have access to the data and the driven analysis.

Of course, care must be taken to make sure there is a good balance between the amount of time and effort that is required to conduct the measurement and that needed for the corresponding analysis of the data measured. An overwhelming measurement strategy is not sustainable in a long haul and will result in an inaccurate and incomplete data set. Therefore, an ideal measurement strategy should have the following characteristics:

- *Simplicity* – it is intuitive and easy to understand.
- *Validity* – it measures what it claims to measure.
- *Robustness* – the measurement value is insensitive to outside factors.
- *Independence* – the measurement values interpretation is not dependent on other measurement values.
- *Prescriptiveness* – the measurement helps project management accurately identify and avoid potential problems (you know which direction is good and which is bad).
- *Analyzability* – the measurement can be analyzed using statistical techniques.

DEFINING METRICS

As previously discussed, one of the first questions to ask in order to establish a measurement strategy is "Why we need to measure?" Of course, the main reason is to produce a quality product. Therefore, there should be clear quality assurance goals and the appropriate measurements to evaluate whether those goals have been achieved. There are several techniques for accomplishing this task. One such technique called *Goal/Question/Metrics* (GQM) [Basili 1984], by which an organization can establish a set of quality goals that need to be achieved. The GQM technique establishes a process to monitor the achievement of the goals.

The GQM technique first requires the organization (or project team) to establish a set of quality goals for the project. It is possible that a large number of quality goals have been identified, and as a result, it may be very difficult, given the project resources and schedule, to achieve all the identified goals. In such a case, the team should prioritize these goals and try to achieve as many of the highest priority goals as possible. Once the quality goals are identified, the team can identify questions about how these goals can be achieved and know when the goals have been achieved. The purpose of these questions is to identify the appropriate metrics that needs to be collected to evaluate whether those goals are achieved. Figure 3.5 represents a GQM analysis.

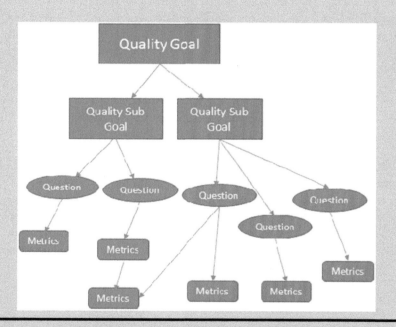

Figure 3.5   GQM analysis.

For example, let us assume that one of our quality goals for the organization is to make sure no more than 5% of the product defects are found after the product release. In this case, we could ask the following questions:

■ During life of the product, how many defects (before and after product release) will be found?
■ What percent of the defects are found before the product release?

- What percent of the defects are found after the product release?
- Which quality assurance activities are best at discovering product defects during the development?

Given the question about percent of defects found before the product release, the following measurements are relevant:

- Number of defects found before the release.
- Number of defects found after the release.
- Number of new defects that are found after the defect fixes.
- Number of defects that are found in the field after the defect fixes.

The four metrics should help answer the question about percent of defects found before the product release. If a quality goal is not achieved, then the organization needs to modify its quality improvement procedures to achieve the quality goal. This may require identifying additional questions, and the corresponding metrics that help achieve the goal, and integrating additional defect prevention and detection activities during the development.

QUALITY METRICS

There are several product metrics that can be used to measure quality; however, generally, they can be grouped into two major categories:

- Quality metrics during the development
- Quality metrics post release

QUALITY METRICS DURING DEVELOPMENT

Software metrics provide insight to the development team about the status of the project, quality of the product, quality of the processes that are used, and provide the basis for decision making. The following are advantages of these metrics:

- Keeping schedules realistic.
- Making project control and risk assessment more quantitative.
- Identifying overly complex programs.
- Identifying areas where tools are needed.
- Providing data for root cause analysis (a method used for identifying the root causes of faults or problems).
- Avoiding premature product release.

Some examples of these metrics include:

- Project metrics: These metrics help the development team to monitor and control the project progress. Some examples of these metrics include:
  - Schedule metrics identify any variation from the project schedule.
  - Cost metrics identify variation from the expected project cost.
  - Requirement change request metrics monitor the number of change requests, their frequency, and when these changes are requested. Unstable requirements are the major cause of project cost and schedule overrun.
  - Effort metrics monitor the project actual progress versus the planned progress.
  - Repair/rework metrics provide information about the quality of the product while under development, and the potential for project schedule and cost overruns.
- Product metrics: These metrics provide the development team insight into the quality of the product under the development.
  - Requirements metrics provide information about the quality of requirements, such as their ambiguity, incompleteness, and complexity. A lower quality requirement points to potential problems throughout the development life cycle.
  - Design metrics provide information about the quality of the design, such as its cyclomatic complexity, cohesion, coupling, function points, depth of inheritance, and stability. The more complex the product design, the higher cost for product maintainability.
  - Implementation metrics provide information about the quality of the code, such as defect density (e.g., defects/KLOC), size of source code (in LOC), size of binary code, code path lengths, and percent of source code compliance with coding standards.
  - Test metrics provide information about product units and the overall product, such as the number of tests cases per software unit, the degree to which test cases cover the requirements, and the number of defects discovered with testing.
- Process metrics: These metrics provides insight about the effectiveness of the processes used to develop a product.
  - Efficiency metrics provide insight about the quality of the process implemented by the development team. For example, the defect removal efficiency provides insight about the quality of the defect removal process, and how many defects were removed before the product release versus the defects found by the end user.

- Cycle time metrics provide insight into the agility of the development team from the time an idea is developed to the time it is in place. For example, cycle time for code development measures from the time design is committed to and then implemented.
- Turnaround time metrics measure the time a defect is identified to the time that defect is fixed.

QUALITY METRICS POST RELEASE

These metrics provide insight into the quality of the product from the customer point of view. Some examples of these metrics include:

- Customer satisfaction
  - Number of system enhancement requests
  - Number of maintenance requests
  - Number of product recalls
- Responsiveness to user requests
- Reliability
  - Availability
  - Mean time between failure
  - Mean time to repair
- Cost of defects to customer
  - Annual maintenance cost
  - Business lost cost
- Ease of use (complexity of the product in operation)

EXERCISE 3.4: DH QUALITY MEASUREMENT STRATEGY

After Massood's discussion of quality measurement, Disha Chandra decided the team needed to create a quality measurement strategy.

The exercise, described at the end of this chapter, deals with the DH Team's discussion.

EXERCISE 3.5: DH GOAL/QUESTION/METRIC

Disha asked the team to use the GQM technique to evaluate the effectiveness of the coding standard being considered for the DH project: *Java Code Conventions* [Sun 1997].

The exercise, described at the end of this chapter, deals with the DH Team's use of the GQM technique.

After Massood's tutorial on SQA, Disha Chandra asked Michel Jackson to prepare and deliver additional material on common quality assurance activities.

MINITUTORIAL 3.4: QUALITY ASSURANCE ACTIVITIES

VALIDATION AND VERIFICATION (V&V)

Validation and Verification are two common terminologies that are associated with quality assurance:

Verification – The process of evaluating a system or component to determine whether the products of a given development phase satisfy the conditions imposed at the start of that phase [IEEE 1012]. In another words, it answers the question of whether the product is being built the right way.

Validation – The process of evaluating a system or component during or at the end of the development process to determine whether it satisfies specified requirements [IEEE 1012]. In another words, it answers the question of whether the right product is being built.

Verification typically involves evaluation of the documents, plans, requirements and design specifications, code units, and test cases. These are sometimes referred to as static evaluation, since no code is executed. On the other hand, validation is the evaluation of the product itself. This is referred to as dynamic evaluation (testing) of the product.

Validation typically follows the verification process. Therefore, the inputs to verification are static products, and its output is the same product with hopefully much higher quality, whereas the input to the validation process is a set of test cases that evaluate the product, and its output is hopefully a perfect product.

There are quality assurance techniques that are associated with V&V, some of these techniques are covered in the following sections.

INDEPENDENT VALIDATION AND VERIFICATION (IV&V)

V&V can either be conducted by the development team or by an independent group which may or may not work in the same organization as the development team; but the independent group is not part of the development team. There are advantages and disadvantages associated with conducting the quality reviews by an independent group. Table 3.2 lists some of the advantages and disadvantages of the IV&V.

Table 3.2   IV&V Advantages and Disadvantages

IV&V Advantages	IV&V Disadvantages
• New set of eyes • Not invested in the development • Not reporting to the same manager • Independent appraisal (potentially higher confidence)	• Additional learning curve • Higher cost

QUALITY REVIEWS

By performing quality reviews throughout the development life cycle we are able to identify and remove the defects as early as possible. As illustrated in Figure 3.2 the earlier a defect if detected, the cheaper it is to fix. Table 3.3 shows the approximate cost/effort associated with the removal of a requirement defect in different phases of the development life cycle.

There are number of different quality review techniques; three such techniques are walk-through, Technical Review, and Inspection [IEEE 1028]. These quality reviews are conducted on different software artifacts, such as requirements specifications, design descriptions, source code, or test plans.

Table 3.3   Cost of Removing a Requirement Defect

Phase	Activities/Resources	Cost
Requirements	Interaction with Stakeholders Requirement documentation update	100–200
Design	Interaction with Stakeholders Requirement documentation update Redesign Design document update	300–1000
Construction and Testing	Interaction with Stakeholders Requirement documentation update Redesign Design document update Coding Testing	1000–10,000

Operation	Interaction with Stakeholders
	Requirement documentation
	  update
	Redesign
	Design document update
	Coding
	Testing
	Repackaging and deployment
		1000–10,000+cost of
		Recall
		Redistribution
		Unsatisfied
		  customers
		Legal (lawyer,
		  litigation, fines, etc.)

Note: Table 3.3 does not consider the activities that are associated with the defect detections and the corresponding cost associated with those activities.

WALK-THROUGH

A *walk-through* is the least formal quality review. During a walk-through, the software artifact owner will lead members of the walk-through team through a review of the artifact. The participants in the walk-through have the opportunity to ask questions and/or make comments about potential problems in the artifact, or potential violation of standards.

There are typically two major objectives for conducting a walk-through:

- To collect input from the walk-through participants about the quality of the product, and whether the product under development is meeting the identified standards.
- To give the participants an opportunity to learn about the product under review; the walk-through is used as a vehicle for knowledge transfer.

The participants in a walk-through are:

- Walk-through Leader – conducts the walk-through, handles the administrative tasks pertaining to the walk-through, and ensures that the walk-through is conducted in an orderly manner.
- Author of the Product – presents the software artifact in the walk-through.
- Recorder – records all decisions and identified actions arising during the walk-through meeting.
- Team Member – prepares for and actively participates in the walk-through, and identifies and describes anomalies in the software product.

TECHNICAL REVIEW

Technical Review is more formal than a typical walk-through. The purpose of a technical review is to evaluate a software artifact using a team of qualified personnel to determine its suitability for its intended use and identify discrepancies from specifications and standards. It provides management with evidence to confirm the technical status of the project. Technical reviews may also provide recommendation and examination of various alternatives, which may require more than one meeting. In a technical review, almost all the participants are technically competent in the same area as the software artifact under review. For example, if the artifact under the review is software design specification, then almost all participants have technical background in software design.

The following are the list of potential participants in a technical review meeting:

- <u>Decision Maker</u> – the person who requested the technical review and is the final authority on whether the meeting has achieved its stated objective.
- <u>Review Leader</u> – performs administrative tasks pertaining to the review, ensures that the review is conducted in an orderly manner, and ensures that the review meets its objectives.
- <u>Recorder</u> – documents anomalies, action items, decisions, and recommendations made by the review team.
- <u>Technical Reviewers</u> – actively participate in the review and evaluation of the software artifact.
- <u>Management Representatives</u> – participate in the technical review to identify issues that require management resolution. (e.g., additional resources needed due to a suggested alternative solution)
- <u>Customer or User Representative</u> – may also participate in the meeting, either for better understanding of the solution approach and/or clarification of its need.

INSPECTION

The purpose of an *inspection* is to detect and identify software product defects. An inspection is a systematic peer examination that does one or more of the following [IEEE 1028]:

- Verifies that the software product satisfies its specifications.
- Verifies that the software product exhibits specified quality attributes.
- Verifies that the software product conforms to applicable regulations, standards, guidelines, plans, specifications, and procedures.
- Collects software engineering data (for example, anomaly and effort data) – data that may be used to improve the inspection (e.g., improved checklists).
- Requests or grants waivers for violation of standards where the adjudication of the type and extent of violations are assigned to the inspection jurisdiction.
- Uses the data as input to project management decisions as appropriate (e.g., to make trade-offs between additional inspections versus additional testing).

Inspection is the most formal quality review technique. The main difference between the inspection and technical review is that in a technical review, the participants may suggest alternative solutions; however, an inspection focuses on detecting and identifying defects. A technical review meeting can be long and laborious, as participants might have long technical discussions about what is the best solution to a specific problem. Inspections meetings are orderly and systematized. This discussion of inspection is based on [IEEE 1028], and the work of Michael Fagan and his "Fagan Inspection" [Fagan 1986].

The following are the list of participants in an inspection meeting:

- Moderator (Inspection Leader) – responsible for planning and organizing the inspection, ensuring the inspection is conducted in an orderly manner and meets its objectives, and ensuring that the inspection data is collected.
- Author – the owner of the document being inspected; contributes to the inspection based on special understanding of the software product, and performs any rework required to make the software product meet its inspection exit criteria.
- Reader – leads the inspection team through the software product in a comprehensive and logical fashion, interpreting sections of the work, and highlights important aspects. This role may be carried out by the author.
- Recorder – documents defects, action items, decisions, waivers, and recommendations made by the inspection team, and records inspection data required for process analysis.
- Inspectors – identify and describe defects in the software product. Only those viewpoints pertinent to the inspection of the product should be presented. In an inspection, everyone also serves as the inspector except the author.

The inspection process is a formal process with number of distinct activities. Figure 3.6 displays the Fagan Inspection process.

The *work product* can be any software artifact (e.g., requirements specification, design description, test plan, source code, etc.). Once the work product is ready for review, then the inspection leader is informed.

For the *preparation,* the moderator identifies the appropriate participants in the inspection process; coordinates with the participants to set up the meeting; collects the work products and all appropriate supporting documents (i.e., references, checklists, inspection logs, etc.); and distributes them to the inspection participants.

The *overview* phase is an optional activity. In cases where the inspectors are not familiar with the work product, the author will provide an overview of the product to make sure everyone has a good understanding of what is being inspected.

During the *individual inspection* phase, each inspector inspects the work product (or the portion assigned). The inspectors record the defects discovered and the time spent inspecting in their individual inspection logs, which are turned into the moderator prior to the inspection meeting.

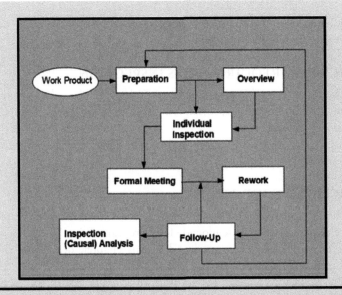

Figure 3.6   Fagan inspection process.

The *formal meeting* is the meeting in which the inspection participants meet to go over the product and identify product defects. Prior to the meeting, the moderator reviews the individual inspection logs to ensure all inspectors have completed their inspection tasks in a proper manner. As the meeting proceeds, the reader goes over the document one item at the time, and the inspectors identify the defects that they have discovered. The recorder records all the defects in the inspection meeting log, and at the end of the meeting, the meeting log is turned into the moderator.

The moderator gives the author a list of all the defects identified during the inspection process. The author will attempt to fix all the defects during the *rework* phase and return the modified work product to the moderator. In some cases, the author may think that an issue identified is a not a defect. In this case, the author will inform the moderator.

Depending on the quality of the product, and the number of defects identified, or other factors, the moderator or some other stakeholder may decide to call for *follow-up* activities. A follow-up activity may be another inspection for the work product, or additional rework by the author.

In some cases, there may be a final phase in the inspection process. This phase is called *inspection analysis* or sometimes referred to as the causal analysis phase. In this phase, the author, selected members of the inspection team, and other stakeholders meet to analyze the type of the defects that were discovered during the inspection phase and look for the root cause for the defects being introduced into the work product, and what can be done to prevent this from occurring again.

EXERCISE 3.6: REQUIREMENTS INSPECTION

After the discussion of Quality Reviews, Disha Chandra asked the team to practice a requirements inspection on the DH High-Level Requirements Definition (HLRD).

The exercise, described at the end of this chapter, deals with the DH Team's inspection.

EXERCISE 3.7: CODE REVIEW

After the discussion of Quality Reviews, Disha Chandra asked the team to practice a code review on pseudocode for a program unit.

The exercise, described at the end of this chapter, deals with the DH Team's code review.

After Michel completed his tutorial on Quality Assurance Activities, Disha Chandra asked Michel "Why didn't you include any material on testing?" Michel said he thought since testing was such a big topic, it would be best to cover it in a separate tutorial. Georgia Magee reminded Disha that she had worked for the last three years as a test engineer at *HomeOwner* and said she would be glad to cover software testing for the team. Disha said "fine – I never turn down a volunteer".

MINITUTORIAL 3.5: SOFTWARE TESTING

The SWEBOK [Bourque 2014] states "*Software testing* consists of the dynamic verification that a program provides expected behaviors on a finite set of test cases, suitably selected from the usually infinite execution domain". Software Testing is a critical element of SQA. It represents the ultimate review of the requirements specification, the design, and the code. It is the most widely used method to insure software quality. Many organizations spend 50% or more of development time in testing.

Key points about software testing:

■ *Testing* is concerned with establishing the presence of program defects, while debugging is concerned with finding where defects occur (in code, design, or requirements) and removing them (fault identification and removal). Even if the best review methods are used (throughout the entire development of software), testing is necessary.

- *Testing* is the one step in software engineering process that could be viewed as destructive rather than constructive: "A successful test is one that breaks the software" [McConnell 2004].
- A successful test is one that uncovers undiscovered defect.
- Testing cannot show the absence of defects, it can only show that software defects are present.
- For most software, exhaustive testing (test all possible cases) is not possible.
- Software tests typically use a *Test Harness*: a system of test drivers, stubs, and other tools to support test execution.
  - A *Test Driver* is a utility program that applies test cases to a component being tested.
  - A *Test Stub* is a temporary, minimal implementation of a component to increase controllability and observability in testing. When testing a unit that references another unit, the unit must either be complete (and tested) or stubs must be created that can be used when executing a test case referencing the other unit.

TESTING TYPES

Figure 3.7 shows the various types of tests associated with the software development life cycle [Pfleeger 2009].

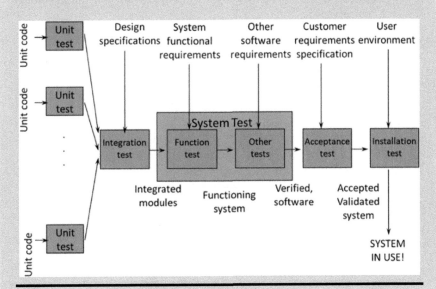

Figure 3.7   Software development tests.

UNIT TESTING

- *Unit Testing* – checks that an individual program unit (function/procedure/ method, object class, package, module) behaves correctly. There are two broad categories of unit testing:
  - *Static Testing* – testing a unit without executing the unit code using a formal review (e.g., performing a symbolic execution of the unit).
  - *Dynamic Testing* – testing a unit by executing a program unit using test data.

There is some debate about what constitutes a "unit". Here are some common definitions of a unit [McConnell 2004]:

- the smallest chunk that can be compiled by itself
- a stand-alone procedure or function
- something so small that it would be developed by a single person

Figure 3.8 illustrates the various techniques used for unit testing [Pressman 2015]. A little detail about some of these techniques:

- Static Techniques
  Symbolic Execution – works well for small programs.
  Program Proving – difficult and costly to apply, and may not guarantee correctness.

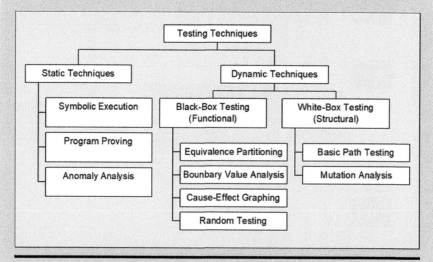

Figure 3.8   Unit testing techniques.

Anomaly Analysis –
- Inexecutable code (island code)
- Uninitialized variables
- Unused variables
- Unstructured loop entry and exit
- Dynamic Techniques

Black Box Techniques –
- Design of software tests relies on module purpose/description to devise test data.
- Uses inputs, functionality, outputs in test design.
- Treats a module like a "black box" (given specific input what would be the expected output?).

White Box Techniques –
- Design of software tests relies on module source code or detailed design to determine test data.
- Treats module like a "white box" or "glass box".

The following issues should be considered when preparing a unit *test plan*:

- Unit test planning is typically performed by the unit developer.
- Planning of a unit test should take place as soon as possible.
- Black box test planning can take place as soon as the functional requirements for a unit are specified.
- White box test planning can take place as soon as the detailed design for a unit is complete.
- If one is testing a single, simple unit that does not interact with other units, then one can write a program that runs the test cases in the test plan. However, if the unit must interact with other units, then it can be difficult to test it in isolation.
  - For example, a unit that loads a system initialization file may require numerous interactions with other units.
  - Such unit testing may require a test harness using drivers and stubs.

INTEGRATION TESTING

The *DigitalHome* development process, defined in Table 3.1, prescribes an incremental construction procedure. As increments/units are developed, they are integrated with the other units previously tested and integrated. As each increment is developed, *integration tests* are conducted to uncover interfacing and interaction errors between the units.

There are various integration testing strategies:

- Bottom-Up – test drivers assist in first testing lower level components.
- Top-Down – component stubs assist in first testing higher level components.
- Sandwich – a mixture of top-down and bottom-up approach.
- Big-Bang – integrate all components together at same time and test them.

Figure 3.9 depicts a set of units developed incrementally, with a test harness of a driver and some stubs. For example, if a top-down approach was used, unit A would be developed first; then followed by integration and testing of units B, C, and D; and then followed by integration and testing of unit E. Such testing can result in significant improvements in finding and correcting interface defects; however, there are added overhead costs.

Typically, a "test scaffold" is built as part of test planning to support such hierarchical testing. Each time a new module is added (or changed), the previous tests of units must be executed again to check for problems not discovered earlier. This is called *regression testing*.

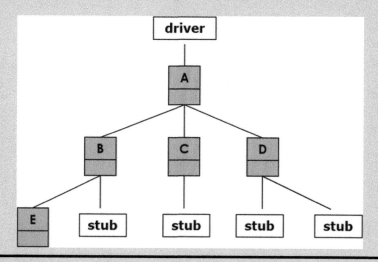

Figure 3.9   Integration testing units.

SYSTEM TESTING

System testing is testing conducted on a complete integrated system to evaluate the system's compliance with its specified requirements.

There are various categories of system testing. Here are some of the categories:

- *Volume Testing* – testing whether the system can cope with large data volumes.
- *Stress Testing* – testing beyond the limits of normal operation (e.g., availability with high loads of use).
- *Endurance Testing* – testing whether the system can operate for long periods?
- *Usability Testing* – testing ease of use.
- *Security Testing* – testing whether the system can withstand security attacks.
- *Performance Testing* – testing the system response time.
- *Reliability Testing* – testing whether the system operates effectively over time.

If use cases (use cases are discussed in Chapter 5) are used to specify the functional requirements for a system, the use cases can be transformed into operational test scenarios (OTS). Each OTS represents the execution of the program for some part of the requirements. Figure 3.10 shows a use case that represents part of the requirements for a system used to manage an airport. The Main Scenario states the requirements associated with use case.

Primary Actor: System Administrator
Goal: Close an airport for specified time interval
Preconditions:
1. System administrator is logged into the system.
2. The system is initialized.
Post Conditions:
1. The airport is closed for the specified time interval.
2. All routes for the airport are closed for the given time interval.
Main Scenario:

Step	Actor Action	Step	System Response
1	The system administrator selects the Close Airport Option	2	The system prompts the user to enter the airport code.
3	The system administrator enters the airport code.	4	The system prompts the user to enter the time interval for closing.
5	The system administrator enters the time interval for which the airport shall be closed.	6	The system closes the airport and displays a confirmation.

Exceptions:
1. The airport entered is a not open during the specified time period.
Response: System displays an error message.
2. System administrator enters an incorrect airport code.
Response: System displays an error message.
3. System administrator enters an incorrect time or incorrect form of time.
Response: System displays an error message.

Figure 3.10　Closed airport use case.

OST Developer: Watts Humphrey   Team _____ X _____ Tester Name _____

Date: 9-26-08   Cycle No. 1   Date _____

Scenario # 10   Scenario Objective: Testing close of an airport in system.

Step	Source	Action	Actual Behavior
1	Program	Execute program and prompt user for initialization text file.	
2	User	Enter file name: "arpinit.txt"	
3	Program	Load file and display main menu. Prompt user for menu item number.	Loads file and displays main menu.
4	User	Select menu option 10 to close an airport.	
5	Program	Prompt user to enter airport ID.	
6	User	Enter "ATL."	
7	Program	Prompt user to enter start time.	
8	User	Enter "0800"	
9	Program	Prompt user to enter end time.	
10	User	Enter "0700"	
11	Program	Print error message for invalid time interval. Prompt user to press enter to continue.	Invalid time interval.
12	User	Press Enter	
13	Program	Prompt user to enter airport ID.	
14	User	Enter "XYZ"	
15	Program	Print error message for invalid airport. Prompt user to press enter to continue.	Invalid airport id.
16	User	Press Enter	
17	Program	Prompt user to enter airport ID.	
18	User	Enter "ATL"	
19	Program	Prompt user to enter start time.	
20	User	Enter "0800"	
21	Program	Prompt user to enter end time.	
22	User	Enter "1000"	
23	Program	Confirm airport is closed between 800 and 1000	Airport closed between 800 -1000
24	Program	Prompt user to press Enter to continue.	
25	User	Press Enter.	
26	Program	Returns to main menu. Prompt user for menu item number.	
27	User	Select menu option 13 to exit program.	
28	Program	Program prompts user yes or no to exit (Y/N).	
29	User	Enter "Y"	
30	Program	Program terminates.	Program terminates with no errors.

Figure 3.11   Closed airport test scenario.

Figure 3.11 shows a test scenario derived from the Close Airport use case. The test scenario describes how the use case test should be carried out, and lists the order of interaction between a user and the system, with the inputs from the user and how the system should react.

For most systems, a special type system test, *an Acceptance Test*, is conducted to see if the system satisfies the customer's requirements. There are two types: of acceptance testing

- <u>Alpha Testing</u> – conducted at the developer's site by the customer.
- <u>Beta Testing</u> – conducted at one or more customer sites by the end users.

Finally, a few key points about software testing:

- Test Plans should be completed as soon as possible:
- System test planning can start during the requirements phase.
- Integration test planning can start during the high-level design phase.
- Unit test planning can start during the detailed design phase.

EXERCISE 3.8: UNIT TEST PLANNING

After the discussion of Unit Testing, Georgia Magee asked the team to study a search algorithm and determine a set of test cases for the unit.

The exercise, described at the end of this chapter, deals with the DH Team's unit test design.

EXERCISE 3.9: INTEGRATION TEST PLANNING

After the discussion of Integration Testing, Georgia Magee asked the team to study a design diagram and determine a strategy for incremental development and integration testing.

The exercise, described at the end of this chapter, deals with the DH Team's unit test design.

EXERCISE 3.9: USE CASE TEST PLANNING

After the discussion of System Testing, Georgia Magee asked the team to study a use case and then develop a test scenario for the use case.

The exercise, described at the end of this chapter, deals with the DH Team's unit test design.

Case Study Exercises

EXERCISE 3.1: SECURITY ATTACKS

After the discussion of the "Security Engineering", Massood lead the team in a discussion about potential attacks on the security of the *DigitalHome* system.

Upon completion of this exercise, students will have increased ability to:

• Describe potential attacks on the security of a software system.
• Understand the importance of building secure software systems.
• Classify the different types of security problems.
• Assess the costs associated with building a secure system.

DH CASE STUDY ARTIFACTS

• DH Development Process Script (Table 2.1)

EXERCISE DESCRIPTION

1. As preparation for the exercise, read Chapter 1 and the Case Study Artifact listed above.
2. You will be assigned to a small development team (like the DH Team).
3. Each team meets, organizes, and carries out the following activities:
   • Each member of team discusses the following items:
     • Security problems they have experienced as a developer or user, or have knowledge in existing systems.
     • Potential security problems they see with the *DigitalHome* system.
     • Whether they have the appropriate level of background and experience to address the security problems.
   • The team identifies the most significant security issues in the team's discussion of their software activities, and summarizes them in a report.
4. Choose one member of the team to present the team's quality report.

EXERCISE 3.2: COST OF QUALITY EXPERIENCES

After the discussion of the "Cost of Quality", Massood lead the team in a discussion of their experiences with the types of costs of quality they had experienced in their previous software development work.

LEARNING OBJECTIVES

Upon completion of this exercise, students will have increased ability to:

• Describe the quality of their software development activities.
• Reflect on the defect life cycle in software they have developed.
• Identify the cost of quality in software they have developed.
• Classify the different types of quality costs.
• Assess whether they have used the appropriate level of resources in their software development activities to achieve the desired level of quality.

DH CASE STUDY ARTIFACTS

• DH Development Process Script (Table 2.1)

EXERCISE DESCRIPTION

1. As preparation for the exercise, read Chapter 1 and the Case Study Artifact listed above.
2. You will be assigned to a small development team (like the DH Team).
3. Each team meets, organizes, and carries out the following activities:
   - Each member of team discusses the following items:
     - The quality of their software development activities.
     - The defect life cycle in software they have developed.
     - The cost of quality in software they have developed.
     - Whether they have used the appropriate level of resources in their software development activities to achieve the desired level of quality.
   - The team identifies the most significant quality issues in the team's discussion of their software activities and summarizes them in a report.
4. Choose one member of the team to present the team's quality report.

EXERCISE 3.3: SOFTWARE QUALITY ASSURANCE PLANNING

After Massood's discussion of the SQA planning, Disha Chandra lead the team in a discussion SQA planning.

LEARNING OBJECTIVES

Upon completion of this exercise, students will have increased ability to:

- Describe the elements of SQA planning.
- Determine the scope and extent of SQA planning needed for a moderate sized project.
- Determine which elements of SQAP, specified in [IEEE 730], are appropriate for *DigitalHome* project.

DH CASE STUDY ARTIFACTS

- DH Development Process Script
- DH HLRD
- IEEE Standard for Software Quality Assurance Processes [IEEE 730]

EXERCISE DESCRIPTION

1. As preparation for the exercise, read Chapter 1 and the Case Study Artifacts listed above.
   You will be assigned to a small development team (like the DH Team).
2. Each team meets, organizes, and carries out the following activities:
   a. Determine which elements of SQAP, specified in [IEEE 730], are appropriate for *DigitalHome* project.
   b. Determine the purpose and scope of the SQA activities.

c. Describe the DH project's SQA organization structure. What are its roles and responsibilities?
d. Decide how problem reports are collected, who will see those data, and what procedures should be followed for implementing the appropriate corrective actions.
e. Choose one member of the team to report to the class on the team's quality planning activity.

EXERCISE 3.4: DH QUALITY MEASUREMENT STRATEGY

After Massood's discussion of quality measurement, Disha Chandra decided the team needed to create a quality measurement strategy.

LEARNING OBJECTIVES

Upon completion of this exercise, students will have increased ability to:

- Understand the value of software measurement.
- Measure software and software development data.
- Analyze and use measured data.

DH CASE STUDY ARTIFACTS

- DH HLRD
- DH Development Process Script

EXERCISE DESCRIPTION

1. As preparation for the exercise, read Chapter 1 and the Case Study Artifacts listed above. You will be assigned to a small development team (like the DH Team). Each team meets, organizes, and carries out the below described activity.
2. The team discusses how to develop a quality measurement strategy and what elements it should contain. As part of this discussion, the team answers the below questions:
   - Why we need to measure?
   - What we need to measure?
   - How to conduct these measurements?
   - What to do with the data collected?
3. Each team answers the questions in writing and chooses one member of their team to report to the class on the group's conclusions.

EXERCISE 3.5: DH GOAL/QUESTION/METRIC

Disha asked the team to use the GQM technique to evaluate the effectiveness of the coding standard being considered for the DH project: *Java Code Conventions* [Sun 1997].

LEARNING OBJECTIVES

Upon completion of this exercise, students will have increased ability to:

- Describe the GQM technique.
- Use the GQM techniques determine the goals, questions, and metrics to solve a problem.
- Explain the values of software metrics and how they may be used.

DH CASE STUDY ARTIFACTS

- DH HLRD
- DH Development Process Script

EXERCISE DESCRIPTION

- As preparation for the exercise, read Chapter 1 and the Case Study Artifacts listed above. You will be assigned to a small development team (like the DH Team). Each team meets, organizes, and carries out the below described activity.
    1. Acting as the DH Team, the team examines the (GQM) method to select metrics that could be used to assess the appropriateness of using *Java Code Conventions* [Sun 1997] as the coding standards for DH code. Determine the following:
        a. A statement of a quality goal (or goals).
        b. Questions about the goal.
        c. Metrics appropriate for answering the questions.
        d. Issues, problems, or costs in collecting the metrics.
    2. Each team describes, in writing, the results of its GQM exercise and reports the results to the class.

EXERCISE 3.6: REQUIREMENTS INSPECTION

After the discussion Quality Reviews, Disha Chandra asked the team to practice a requirements inspection on the DH HLRD.

LEARNING OBJECTIVES

Upon completion of this exercise, students will have increased ability to:

- Organize and perform a requirements specification inspection.
- Appreciate the value of an inspection.
- Develop correct requirements statements.

DH CASE STUDY ARTIFACTS

- DH HLRD
- DH Development Process Script

EXERCISE DESCRIPTION

- As preparation for the exercise, read Chapter 1 and the Case Study Artifacts listed above. You will be assigned to a small development team (like the DH Team). Each team meets, organizes, and carries out the below described activity.
  1. Acting as the DH Team, the team organizes as an inspection team: roles are assigned (e.g., moderator, recorder, etc.); material is prepared for recording defect and inspection time. Then, the team carries out the following activities:
     a. Inspectors inspect the HLRD in Chapter 1, and record defects found, and time spent.
     b. The inspection team meets and discusses the defects found.
     c. The team agrees on defects in the product and discusses needed product rework.
     d. The team prepares a report on the inspection.
  2. One of the team member reports to the class on the team's inspection.

EXERCISE 3.7: CODE REVIEW

After the discussion of Quality Reviews, Disha Chandra asked the team to practice a code review on pseudocode for a program unit.

LEARNING OBJECTIVES

Upon completion of this exercise, students will have increased ability to:

- Perform a code review of a software code unit.
- Understand the value of a careful code review.
- Use code review data to improve reviewing effectiveness.

DH CASE STUDY ARTIFACTS

- DH Development Process Script

EXERCISE DESCRIPTION

- As preparation for the exercise, read Chapter 1 and the Case Study Artifacts listed above. You will be assigned to a small development team (like the DH Team). Each team meets, organizes, and carries out the below described activity.
  1. Acting as the DH Team, each member of the team acts as code reviewer. Then, each reviewer carries out the following activities:
     a. Reviews the below **gcd** (greatest common divisor) procedure, and records defects found and time spent.
     b. The team meets and compares what reviewers found.
     c. The team agrees on defects in the procedure and discusses needed product rework.
     d. The team prepares a report on the code review.
  2. One of the team members reports to the class on the team's code review.

```
procedure gcd (int a, int b, int value) {
  // author: Euclid
  int temp, value;
  1. a := abs(a)
  2. b := abs(b)
  3. If (a=0 and b=0) then
  4.     raise exception
  5. else if (a ≠ 0) then
  6.     value := b;   // b is the GCD
  7. else
  8.     while (b ≠ 0 ) loop
  9.         temp := b;
  10.        b := b mod a;
  11.        a := temp;
  12.    end loop
  13.    value := a;
  14. end if;
  15. end gcd
```

EXERCISE 3.8: UNIT TEST PLANNING

After the discussion of Unit Testing, Georgia Magee asked the team to study a search algorithm and determine a set of test cases for the unit.

LEARNING OBJECTIVES

Upon completion of this exercise, students will have increased ability to:

- Determine a set of test cases for a software unit.
- Plan for effective testing of a software unit.
- Understand the differences between white-box testing and black-box testing.

DH CASE STUDY ARTIFACTS

- DH Development Process Script

EXERCISE DESCRIPTION

- As preparation for the exercise, read Chapter 1 and the Case Study Artifacts listed above. You will be assigned to a small development team (like the DH Team). Each team meets, organizes, and carries out the below described activity.
  1. Acting as the DH Team, carries out the following activities:
     a. Each member reviews the below binarySearch procedure and develops a set of test cases for the procedure.
     b. The team meets and compares what the team members developed. There is a discussion of which test cases are related to black-box testing and which to white-box testing.
     c. The team agrees on a set of test cases to be used for testing the unit.
     d. The team prepares a report on unit test planning.
  2. One of the team members reports to the class on the team's unit test planning.

```
function binarySearch(List, int n, int value):
// purpose: search List (an array of n
//   integers) to find the position m of value
//   in List
1. left := 0
2. right := n – 1
3. while (left < right) loop
4.     m := int(left + right)/2
5.     if List[m] < value:
6.         left := m + 1
7.     else if List[m] > value:
8.         right := m - 1
9.     else:
10.        return m
11. return unsuccessful
12. end binarySearch
```

EXERCISE 3.9: INTEGRATION TEST PLANNING

After the discussion of Integration Testing, Georgia Magee asked the team to study a design diagram and determine a strategy for incremental development and integration testing.

LEARNING OBJECTIVES

Upon completion of this exercise, students will have increased ability to:

- Understand incremental development.
- Determine a strategy for integration testing.
- Understand the differences between top-down and bottom-up integration testing.

DH CASE STUDY ARTIFACTS

- DH Development Process Script

EXERCISE DESCRIPTION

- As preparation for the exercise, read Chapter 1 and the Case Study Artifacts listed above. You will be assigned to a small development team (like the DH Team). Each team meets, organizes, and carries out the below described activity.
    1. Acting as the DH Team, carries out the following activities:
        a. Each member reviews the below *DigitalHome* Conceptual Design Diagram (Figure 2.11 in Chapter 2).

b. The team meets and discusses in what order the modules/units will be developed and integrated. The team will decide on whether to implement a top-down or a bottom-up integration testing strategy.

c. The team agrees on it strategy.

d. The team prepares a report on integration testing.

2. One of the team members reports to the class on the team's discussion and decisions.

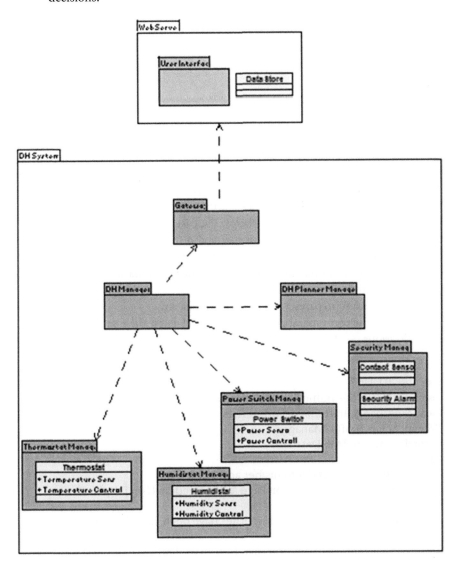

EXERCISE 3.10: USE CASE TEST PLANNING

After the discussion of System Testing, Georgia Magee asked the team to study a use case and then develop a test scenario for the use case.

LEARNING OBJECTIVES

Upon completion of this exercise, students will have increased ability to:

- Understand system test planning.
- Understand the structure of a use case.
- Develop a test scenario from a use case.

DH CASE STUDY ARTIFACTS

- DH Development Process Script

EXERCISE DESCRIPTION

- As preparation for the exercise, read Chapter 1 and the Case Study Artifacts listed above. You will be assigned to a small development team (like the DH Team). Each team meets, organizes, and carries out the below described activity.
  1. Acting as the DH Team, carries out the following activities:
     a. Each member reviews the below use case "Set Traffic Signal Properties".
     b. Each member develops a test scenario (like the one in Figure 3.11) for the Set Traffic Signal Properties use case.
     c. The team meets and agrees on a team test scenario for the Set Traffic Signal Properties use case.
     d. The team prepares a report on the scenario.
  2. One of the team members reports to the class on the team's discussion and decisions.

Use Case ID: UC3
Use Case Name: Set Traffic Signal Properties
Goal: Set/Change traffic signal properties
Primary Actors: Traffic Manager
Secondary Actors: TMS Database
Pre
Traffic Manager is logged into his/her account.
TMS Traffic Signal Configuration Page is displayed on User Display Device.
Post
All newly entered traffic signal properties have been stored in the TMS Database.
TMS Traffic Signal Configuration Page is displayed on User Display Device.
Main Success Scenario

Step	Actor Action	Step	System Reaction
1	Select "Set Traffic Signal Properties".	2	Display "Select the traffic signal(s) for which you wish to set properties".
3	Select one or more traffic signals.	4	Display "Enter Green Light duration (in seconds)".
5	Enter seconds of Green Light duration.	6	Store Green Light duration for selected traffic signals in the TMS database.
		7	Display "Enter Red Light duration (in seconds)".
8	Enter seconds of Red Light duration.	9	Store Red Light duration for selected traffic signals in the TMS database.
		10	Display "Enter Amber Light duration (in seconds)".
11	Enter seconds of Amber Light duration.	10	Store Amber Light duration for selected traffic lights in the TMS database.
		12	Display "Do you wish to set the properties for other traffic signals? YES/NO?"
13A	Enter "YES".	14A	Go to Step 1.
13B	Enter "NO".	14B	Display TMS Traffic Signal Configuration Page.

UC GUIs: DH Configuration Page, DH Default Parameter Configuration Page
Exceptions (Alternative Scenarios of Failure):
At any time, User fails to make an entry.
   a. Time out after five minutes and log out from the system.
System detects a failure (loss of Internet connection, power loss, hardware failure, etc.).
   b. Display an error message on the User Display Device.
   c. System restores system data from TMS Database.
   d. Log off user.
Use Cases Utilized: None
Notes and Issues: The duration of individual traffic signal states must be in the following duration ranges: Red [30–120 seconds], Green [30–120 seconds], and Amber [5–10 seconds]

Chapter 4

Managing the DH Project

Project Planning

In mid-September 202X, a two-day workshop meeting was held in the conference room of the office of the *DigitalHomeOwner* Division of *HomeOwner* to launch the *DigitalHome* (DH) project. Prior to the workshop, the DH Project Team (Disha Chandra, Michel Jackson, Yao Wang, Georgia Magee, and Massood Zewail) had decided on team roles and responsibilities had been assigned. During the launch workshop, the team analyzed the DH Need Statement, established a Development Process, created a Conceptual Design, and determined a Development Strategy. (See Chapter 2 for additional detail on the project launch.)

At the end of the workshop, Disha Chandra scheduled a project planning meeting for the next week and asked Massood Zewail, the DH Planning Manager, to lead the meeting and the project planning activities.

MINITUTORIAL 4.1:   PROJECT MANAGEMENT OVERVIEW

In the SWEBOK [Bourque 2014], Software Engineering Management is defined as "the application of management activities—planning, coordinating, measuring, monitoring, controlling, and reporting—to ensure that the development and maintenance of software is systematic, disciplined, and quantified". The type and extent of project management activities used depends on the size and complexity of the project and the nature of the software process.

For a small project using an agile process, management activities might concentrate on regular team meetings, and planning and scheduling increments; and the effort devoted to planning, measuring, monitoring, and reporting might be minimal. Larger projects, using heavyweight plan-driven processes, would likely have well-defined activities for software estimation, task planning and scheduling, risk management, configuration management, and quality planning.

Many elements of software project management are similar to the management of the development of any type of product; however, the special nature of software (see the discussion of complexity, changeability, conformance, and invisibility in Chapter 1) and the accompanying cost and quality issues associated with the development of large, complex software systems requires special attention to activities such as estimation, measurement, monitoring, and to risk, change, and quality management. Because of this there has been greater emphasis on software management processes, procedures, and standards. For example, the CMMI (Capability Maturity Model Integration) [SEI 2010] (see Chapter 2) lists the following key process areas, which are directly related to project management:

- Project Planning
- Project Monitoring and Control
- Supplier Agreement Management

- Integrated Project Management
- Risk Management
- Quantitative Project Management
- Configuration Management
- Measurement and Analysis

In a similar vein, the IEEE Standard on *Systems and Software Engineering-Software Life Cycle Processes* [IEEE 12207] lists the following processes for project management: Project Planning, Project Assessment and Control, Decision Management, Risk Management, Configuration Management, and Measurement and Analysis.

Table 4.1 describes some of the key activities that are part of software project management (based on IEEE 12207 and the SWEBOK)

Table 4.1   Project Management Activities

Activity	Description
Project Planning	Establish project feasibility; determine the project scope; determine project deliverables; identify project tasks; estimate product size and required project effort and duration; establish schedules and responsibility for project task completion; create a measurement process, deciding what measurement data will be collected and how the data will be used; and decide on required resources to accomplish project tasks (staffing, training, and tools, equipment and infrastructure).
Project Monitoring & Control	Determine the status of the project and ensure that the project performs according to plans and schedules, within projected budgets, and that it satisfies technical objectives; and redirect project activities, as appropriate, to correct identified deviations and variations from project plans and objectives.
Risk Management	Identify the risks to accomplishing the project objectives; analyze the risks to determine their likelihood of occurrence and their potential impact; establish a risk management plan; and monitor and manage the risks.

(Continued)

Table 4.1 (*Continued*) Project Management Activities

Activity	Description
Configuration Management	Manage and plan the SCM (Software Configuration Management) process; identify software configuration items and establish a configuration baseline; manage software change control; execute configuration status accounting and configuration auditing; and manage software release and delivery.
Quality Planning	Determine the objectives for product and process quality; identify standards, methodologies, procedures, metrics, and tools for assuring the quality of the software product and processes; set quality metric goals for project deliverables and the verification and validation activities specified in the project task plan; and plan for software assurance and auditing activities. (See Chapter 3 for additional detail about software quality.)

A large, complex "plan-driven" project would likely engage in all of the management activities described in Table 4.1. However, a project using an "agile" process (see Chapter 2 for detail about plan-driven and agile processes) would limit its full use of these activities. For example, eXtreme Programming (XP) [Beck 2000] has a practice call the "Planning Game", which involves quickly determining the scope of the each release, with the customer deciding on the scope, priority, and release dates from a business perspective, and technical people estimating and tracking progress. The focus would be on doing detailed planning (scope, priorities, cost, and schedule) for the next release, while a less-detailed long range plan would be refined after each release.

How a project team is structured is one of the early decisions that must be made in the project management process. Project structure may depend on number of factors – the project domain (e.g., health, transportation, entertainment, finance), the type of application (e.g., shrink-wrapped, real-time embedded, web-based, etc.), the size and complexity of the project, and the nature of the project's parent organization. Three common ways to classify project structures are illustrated in Table 4.2.

Table 4.2 Project Organizational Structures

Type	Description
Project Structure	Personnel are organized into teams that are responsible for carrying out all activities associated with a particular project.
Functional Structure	Personnel are organized into teams according to a particular software engineering function (e.g., requirements, design, construction, quality, etc.).

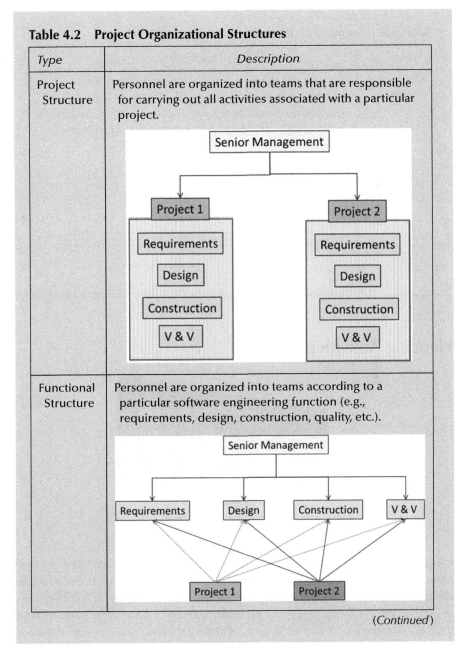

(Continued)

Table 4.2 (*Continued*)   Project Organizational Structures

Type	Description
Matrix Structure	The project structure and functional structure are combined using aspects of both; the functional teams provide resources to the project teams (e.g., a Design team carries out design activities for all project teams).

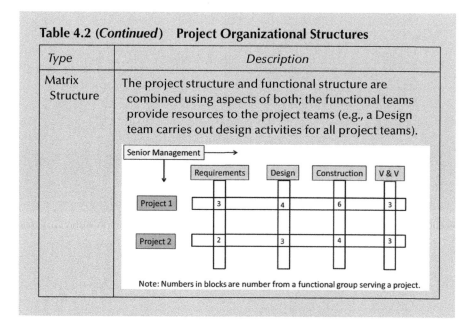

Note: Numbers in blocks are number from a functional group serving a project.

Planning Activities

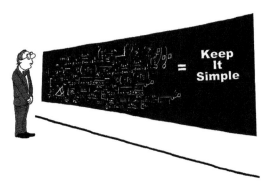

During one of the breaks in the mid-September 202X two-day *DigitalHome* Launch Workshop, Disha Chandra and Massood Zewail had a short conversation about *DigitalHome* development activities (see Chapter 2). Massood had been designated as the Planning Manager and Disha wanted to know how he wanted to proceed. Massood said that he is torn between following the DH Development Process (Table 2.2), which the team had just come up with and has been used successfully in industry, or follow an agile process, say Scrum, which has become popular in industry. He took a Software Project Management course in his graduate program and felt he understood the basics of the plan-driven process and agile processes; but, he felt a bit uncomfortable about leading the effort because he had the least experience of anybody on the team and now he had an added responsibility of delivering series of MiniTutorials about the Scrum process (see Chapter 10). Disha tried to reassure him by pointing out that "some of us have had a lot of experience dealing with poor plans or poorly executed plans" and that she wanted this to be a team effort, with Massood acting as a facilitator. Also, she felt Massood, because of his graduate work, might be more current about recent project planning methods and approaches.

She asked Massood to prepare for a planning meeting she would schedule in the following week.

MINITUTORIAL 4.2: SOFTWARE ESTIMATION

Unfortunately, software projects have a history of budget and schedule problems: poor budget/schedule planning, excessive and unexpected costs, and missing schedule milestones. Part of this is due to the nature of software and the dynamic nature of software engineering methods and tools (discussed in Chapter 1). However, a significant contribution to software budget and schedule problems is a lack of good estimates of the size of the software product and the effort required to produce it.

Other areas of engineering do not seem to have the same type of problem. For example, in the planning for office building construction, estimates of size and cost are pretty straight forward: the basic size measure of square feet is comparatively easy to determine, depending on the intended use of the building; and the cost per square foot is based on historical data about similar structures. Of course, refinements must be made, based on such things as the number of floors, construction regulations, and varying costs of material and labor.

Three fundamental principles of estimation [Fairley 2009] that might guide and inform project estimation are as follows:

- *Principle 1* – A project estimate is a projection from past experiences to the future, adjusted to account for differences between past and future.
- *Principle 2* – All estimates are based on a set of assumptions that must be realized and a set of constraints that must be satisfied.
- *Principle 3* – Projects must be re-estimated periodically as understanding grows and as project parameters change.

SIZE ESTIMATES

Project cost and duration are two essential estimation values that are needed for project planning. From these estimated parameters other estimates related to staffing and the schedule can be derived. Cost is typically, at least initially, estimated in terms of person-effort units (hours, days, months). Effort is often estimated based on the size of the product to be developed. For example, suppose one estimates the size of a software product to be 10,000 lines of code (10,000 LOC or 10 KLOC). Then, using historical data that says a developer can, on average, produce 2.5 LOC/hour, it could be estimated that the product will take (10,000 LOC) ÷ (2.5 LOC/hour) = 4000 hours of effort to produce or about two person-years of effort. If the product could be divided into manageable development parts, this might translate into a 24 person three-month project.

There are various size measures used: LOC (non-comment lines of source code), object classes, use cases, GUI pages, and function points. A *function point* is a numerical unit determined from estimates of the "external size" of the product, by combining various measures of system functionality (e.g., number of outputs, inquiries, inputs, internal files, and external interfaces).

Most estimation models use either function points or LOC as the size measure. The advantage of using function points to estimate size is that they are easier to determine than LOC from the functional requirements; LOC are difficult to estimate unless one has some concept of the system design. However, once a system is complete, LOC are easier to count, which is important if the estimate is to be checked for accuracy and if historical data are to be collected. Also, there are techniques for converting function points to LOC.

ESTIMATING TECHNIQUES

Expert Judgment

An expert's background and experience in a particular application domain can provide a foundation of historical data (in the mind of the expert) and the insight to make judgments about the effort and duration of a project. A person with great experience in software projects of the type being considered can make estimates based on both the functional and non-functional requirements of the system being developed. However, care must be taken when the system being developed is not similar to the expert's specialty area, or where the expert exhibits certain preconceptions or biases.

The Delphi Method

The Delphi method, a special form of expert judgment, is a systematic, controlled estimation technique, which relies on a group of experts (wise men) to answer questionnaires, anonymously, in two or more rounds. After each round, a facilitator provides an overview of the experts' estimates, as well as the reasons for the various judgments. Based on this information the experts may modify their judgments in the next round. The revision of the individual estimates over multiple rounds typically leads to a narrowing of the gap between estimates, and results in a "group estimate". As might be expected, the validity of the method depends on the quality of the experts and the capability of the facilitator.

Estimation by Analogy

Estimation by analogy is a way of estimating planning information for a new project by using historical data about similar completed projects. If good information on the analogous project is available and the analogy is strong, this technique can provide a reasonable estimate in a speedy fashion. For example, if a new project involved the development of an online reservations system for a large hotel chain, an analogy might be drawn to the development of a similar system for a motel chain. If the motel reservation system used 3650 hours of effort by four software engineers over a six-month time period to develop the system and the scope and features of the hotel and motel reservation systems were similar, an initial estimate for the hotel system might start with the motel system data.

Since no two software projects are identical, adjustments would need to be made based on project differences. This is where another technique, such as expert judgment or the Delphi method, might be employed.

Business-Driven Techniques

Two techniques that are related to business matters are as follows:

- *Pricing-to-Win* relates to the submission of bids to win a contract. Estimates are tailored so that the business is likely to win the bid and make a profit or secure some other value (such as set a foundation for future business).
- *Estimating to Available Capacity* is a technique where estimates of effort, staffing, and duration are shaped to the business's current capacity to accommodate these resources.

Statistical/Parametric Methods

This approach involves building an estimation model from historical data for previously completed projects. The model would typically consist of a parametric equation: $E = f(P)$, where E = estimated value (e.g., effort, duration, number of developers) and P = estimated parameters (e.g., LOC, function points, effort, project characteristics)

Figure 4.1 is an example of the use of a statistical estimation model. The table in the figure lists historical data for ten programs – the LOC for each program

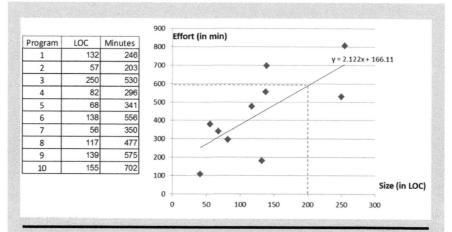

Program	LOC	Minutes
1	132	246
2	57	203
3	250	530
4	82	296
5	68	341
6	138	556
7	56	350
8	117	477
9	139	575
10	155	702

Figure 4.1   Linear regression estimation.

and the number of minutes for development of each program. (This is data for development of the ten programs specified in [Humphrey 1995].) The ten sets of values are plotted on Minutes versus LOC axes. The straight line and its equation ($y=2.122x+166.11$) were determined by a "linear regression" analysis of the table values. Linear regression is a statistical method that models the data to the straight line that best fits all the points (i.e., comes closest to all points, on average). If the analysis is "valid" (the sample data set is sufficiently large) and the data "correlates" (the line is a good fit), the line can be used to predict effort (in minutes) for a given size estimate (in LOC). For example, if it is estimated a new program to be written will have 200 LOC, the predicted effort would be $y=2.122\times200+166.11=591$ minutes or about 9.8 hours.

Of course, a good estimation model must be supported by historical data that is sufficient and relevant. It would be influenced by attributes such as application type and domain, developer experience, system characteristics (such as complexity, reliability and efficiency), the development infrastructure, and the operational environment. Also, it should be noted that the regression model may not be linear; but, rather it could represent some other mathematical relationship, such as a quadratic or exponential function.

There are various established techniques for estimation that are in common use for estimating software project effort, schedule and costs; two of the most popular approaches are COCOMO (Constructive Cost Model) [Boehm 2000] and SLIM (Software Life cycle Management) [Fairley 2009].

COCOMO is a widely used software cost estimation model, developed by Barry Boehm [Boehm 2000], which uses a basic regression formula with parameters that are derived from historical project data and project attributes. There are several different levels of COCOMO that depend on available information about project attributes, their impact on the project estimates, and whether more detailed estimates about various phases of the project (analysis, design, etc.) are desired. The basic COCOMO model uses the below parametric equations:

$$E = \text{Effort}\left(\text{in person-months}\right) = a\left(\text{estimated KLOC}\right)^{b}$$

$$D = \text{duration}\left(\text{in months}\right) = c\left(E\right)^{d}$$

Where the parameters a, b, c, and d depend on the size and type of project.

The SLIM estimating technique uses a theory-based approach based on work of Larry Putnam [Putnam 1991]. The details of SLIM are beyond the scope of this tutorial; but the technique basically involves the modeling of project estimates for effort (E) and duration (D). The model provides a set of non-linear simultaneous equation with unknowns of E and D and input parameters related to estimation of project characteristics such as size, productivity, and the rate of staffing changes. Then numerical methods are used to solve the equations for E and D.

There are excellent tools available for making both COCOMO and SLIM estimates.

MINITUTORIAL 4.3: PROJECT PLANNING

Project planning involves the set of activities needed to develop a project plan: determining the scope of the project; choosing the appropriate software development process; deciding on the methods and tools to be used; agreeing on the project deliverables; decomposing project work into a set of tasks to be competed; estimate project effort, staffing, and duration (discussed in MiniTutorial 4.2); and scheduling the tasks over the estimated duration of the project.

WORK BREAKDOWN STRUCTURE

Some of the planning elements may already be determined prior to the beginning of the project planning activities; for example, the project process and duration, and any methods and tools mandated by organizational standards or business concerns. Also, the selection of a process, say plan-driven or agile, can affect the tasks to be performed and the number and type of deliverables. A widely used approach to identifying project tasks is to develop a Work-Breakdown-Structure (WBS). A WBS is hierarchical: it decomposes project work into a set of task components, which in turn can be decomposed into smaller components. The lowest level components would represent tasks that could be assigned responsibility and be associated with deliverables and specific estimates of size, effort, and duration.

Figure 4.2 pictures a WBS for a Traffic Management System (TMS) – a system that manages and controls the traffic in a medium size city (e.g., 100,000 population). Only one complete thread of decomposition is shown: TMS → Requirements Engineering → Analysis → Context Diagram → Use Case Model → Class Diagrams → Sequence Diagrams. The lowest level components represent tasks associated with the development of Requirements Analysis models.

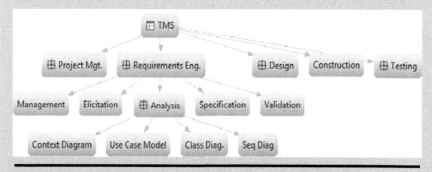

Figure 4.2   Traffic management WBS.

PROJECT SCHEDULING

Once project tasks and deliverables are determined a schedule for task and deliverable completion can be developed. There are a number of tools and techniques used for developing and presenting project schedules: the Critical-Path Method (CPM), the PERT (Program Evaluation and Review Technique) Network, the Gantt Chart, or simply a customized table or spreadsheet.

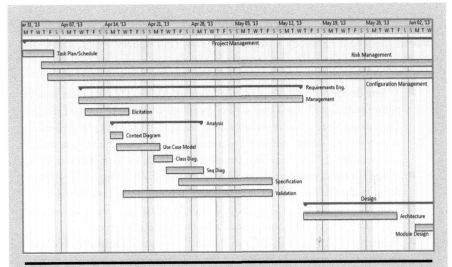

Figure 4.3   TMS Gantt chart

Figure 4.3 shows a portion of the schedule for TMS work using a Gantt Chart (sometimes called a Task-Gantt chart). The Tasks in this chart are drawn from the TMS WBS in Figure 4.2. The horizontal bar represents a time line for the planned completion of a task, starting on the left with the planned start date and ending on the right with the planned finish date. The thin black lines represent tasks with a set of subtasks. The wider gray bars are for lower-level tasks. For example, the Requirements Engineering time line has subtask time lines for Requirements Management, Elicitation, Analysis, Specification, and Validation. Notice that the Analysis time line is further subdivided into time lines for the Context Diagram, Use Case Model, Class Diagrams, and Sequence Diagrams tasks.

Tools that generate Gantt Charts, like Figure 4.3, typically allow for other schedule-related activities: assigning task responsibility to a project team member; describing resources need for task completion; estimating expected task effort and recording actual task effort; monitoring the progress on task completion; designating project milestones, and recording their achievement; and indicating the current state of the schedule. For instance, the vertical orange-yellow line in Figure 4.3 represents the state of the project on a particular date (5/02/2013 in Figure 4.3).

EARNED VALUE TRACKING

Monitoring a project schedule involves regular checks on plan progress: recording planned effort versus actual effort, checking on task completion, and noting milestone achievement. Earned value tracking is a technique for monitoring the progress on a project task plan [Humphrey 1995]. The technique consists of the following steps:

- When developing a task plan, the Planned Value (PV) for each task is computed. The PV for a task is the percentage of total planned effort that the task planned effort represents. For example, if the total planned effort for a project is 4800 hours and planned effort for the requirements elicitation task is 120 hours, then the task $PV = (120/4800) \times 100\% = 2.5\%$.
- When a task is completed, an Earned Value (EV) is logged. The EV is equal to the PV for the task. There is no EV for partially completed tasks.
- Project progress can be judged by looking at the Cumulative $EV(T) = \sum PV$, overall completed tasks completed by time T.
- Figure 4.4 shows a plot of Cumulative PV versus Cumulative EV over the life of a project, 11 months. The PV curve shows a steady rate of effort over a planned ten months to complete the project. The EV curve, representing the actual rate of progress (% of task completion), shows a more irregular completion of tasks and 11 months to complete the planned ten-month project.

A nice feature of the Cumulative EV is that it provides a simple measure, at any point in the project, of whether the project is ahead or behind schedule, and by how much, in terms of task completion. Such tracking provides the opportunity to make adjustments in project work to get back on schedule, or to renegotiate over requirements, deliverables, or the delivery date of a product.

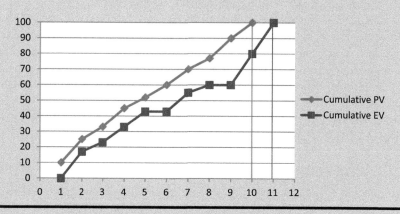

Figure 4.4   Cumulative PV versus cumulative EV.

EXERCISE 4.1: MAKING A TASK PLAN AND SCHEDULE

Prior to the project planning meeting, Massood asked for a meeting with Disha and Jose Ortiz, the Director of the *DigitalHomeOwner* Division, to discuss the type and extent of planning activities envisioned. Disha reiterated the points in the planning section of the Project Development Process: developing Task, Schedule, and Resource Plans, a Risk Plan, a Quality Plan, and a Configuration Management Plan. She suggested they start with task planning and scheduling, and leave risk, configuration management, and quality planning until later.

Massood mentioned using COCOMO to make estimates and developing a WBS, a PERT Network, and a Gantt Chart. Jose cautioned that they should keep things simple: start with the team making approximations of size and effort (using their own experiences with productivity) and then develop a simple WBS diagram of the DH component tasks, and a spreadsheet with a task plan and schedule; he felt this would be sufficient for beginning of the project.

The exercise, described at the end of this chapter, deals with the DH Team's efforts in developing a task plan and schedule, with Massood facilitating.

Risk Management

After completion of the DH task plan and schedule, Massood led the team in a discussion of risk management. There were questions and discussion about what kinds of risks might be encountered, how they would be analyzed, and what action should be taken to reduce the likelihood and impact of the risks.

MINITUTORIAL 4.4: RISK MANAGEMENT

RISK IDENTIFICATION

One of the first tasks in managing risks is to identify and classify potential risks. Various identification techniques are used [Fairley 2009, Sommerville 2011], including use of a checklist, brainstorming, expert judgment, and trade-off analysis. Likewise there are a variety of risk classification systems: for example, by using risk factors such as personnel, estimation and schedule, requirements, process, technology, quality, and other factors that involve risk. In this discussion, we will classify risks as follows:

- *Project Risks* – risks related to the project personnel, schedule, or resources. (e.g., the TMS project team may have a risk related to whether sufficient time has been scheduled for system testing.)
- *Technical Risks* – risks related to the techniques, procedures, methods, tools, and other technical factors affecting the development of a product or service. (e.g., if a TMS adopts a new design methodology, there are risks related to the time for the project team to learn the methodology, proper application of the methodology, and the quality of the resulting design). This category of risks also includes security threats (e.g., an unauthorized user logs on to the system successfully).
- *Business Risks* – risks related to a business or organization that acts as a stakeholder for the product or service being developed. (e.g., Suppose *TransCompute* is the company that won the TMS development contract and is responsible for TMS development. If the contract is fixed-cost, there are risks associated with completing the project with sufficient quality to retain the company's reputation for excellent work and with a return on investment that ensures its viability.)

However, a risk can overlap several of the categories. For instance, the employment of a new design methodology by the TMS project team could introduce an above average number of defects in the TMS software (a technology risk), which could in turn increase the cost and time to perform system testing (a project risk). This could result in financial losses for *TransCompute* because of the fixed-cost TMS contract (a business risk).

RISK ANALYSIS

After risks are identified, the first step in Risk Analysis is to assess how likely is it that a particular risk occurs, and if the risk does occur, what is its impact on successful completion of the project (reaching its cost, schedule, and quality goals). This combination of risk likelihood and risk impact is referred to as *risk exposure*.

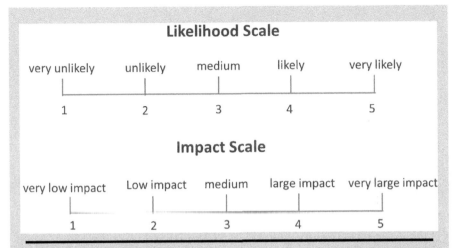

Figure 4.5   Risk exposure scales.

Risk likelihood is expressed as a probability. A quantitative expression of likelihood would be a number between 0 and 1 (or from 0% to 100%). However, such precision may be difficult to determine, so a qualitative likelihood scale can be used; for example, Figure 4.5 has discrete scale from 1 (very unlikely) to 5 (very likely). Suppose "insufficient time is scheduled for systems testing" has been identified as a risk for the TMS project. In her risk analysis of the TMS project, the *TransCompute* project manager, who has extensive experience with projects of the size and scope of TMS, estimates the likelihood of this risk occurring as 2 (unlikely).

In estimating the impact if a risk occurs, various measures could be used: increased cost or effort, extended schedule, business reputation, etc. Because of such variation in measurement of impact, a qualitative approach using a scale such as in Figure 4.5 is most appropriate.

To perform a risk analysis for the TMS project the project manager might lead the team in carrying out the following activities:

■ Identity the risks and describe them.
■ Estimate, for each risk, the occurrence likelihood.
■ Estimate, for each risk, the impact if the risk occurs.
■ Compute the Rick Exposure for each risk (the product of likelihood and impact).
■ Order the risks from highest risk exposure to the lowest.

Table 4.3 shows the results of the TMS risk analysis, with the risks ordered by Risk Exposure. Risk 102 has the highest exposure at 16. One might ask how the Likelihood and Impact of such a risk might be determined. One common approach is expert opinion. The project manager or some other member of the project team might have extensive experience with system testing on past *TransCompute* projects. If many past projects had required extensive testing because of numerous test defects and such problems had caused schedule and budget overruns, Likelihood and Impact would be rated high. Also, if *TransCompute* had a robust metrics program, quantitative data analysis might be used to arrive at the risk exposure factors.

Table 4.3   TMS Risk Analysis

ID	Description	Type	Likelihood	Impact	Risk Exposure
102	The schedule for system testing is underestimated.	Project	Likely (4)	Large (4)	16
111	*TransCompute* loses money on the TMS project.	Business	Medium (3)	Large (4)	12
103	Problems due to new design methodology.	Technical	Unlikely (2)	Very large (5)	10
101	Loss of a key member of the project team at critical point.	Project	Unlikely (2)	Large (4)	8
109	Difficulties interfacing with TMS sensors and controllers.	Technical	Unlikely (2)	Large (4)	8
104	City is unhappy with TMS implementation.	Business	Very unlikely (1)	Very large (5)	5
110	TMS implementation is delayed by local calamity (e.g., a hurricane)	Business	Very unlikely (1)	Medium (3)	3
...	...	...	...	...	...

RISK PLANNING AND MONITORING

Risk identification and analysis are the first steps in planning how to manage the risks. Risks can be managed in a number of ways:

- *Avoidance* – Avoid high risks
  - E.g., if the risk of a new design methodology seems too high (as in Risk 103 in Table 4.3), management may delay the adoption of a new methodology, instead employing one the team is more familiar with. Such a decision might be coupled with tryout of the new design methodology in a pilot project, in order to test its viability.
- *Control* – Use traditional project management techniques to control risks
  - E.g., for Risk 102 in Table 4.3 the project could increase and improve the pre-test quality techniques: reviews, and inspections, collection of quality metrics, test-driven design, and improved documentation.
- *Assumption* – If potential benefits are high enough and probability of risk occurrence is low, accept the risks
  - E.g., Risk 111 in Table 4.3 may be accepted as unavoidable and uncontrollable. It may not involve a direct loss of revenue (depending on the terms of the contract); however, it might delay the start of other projects because of limited available resources.
- *Transfer* – If a risk seems high in one area, transfer it another area
  - E.g., For Risk 109 in Table 4.3, a decision to subcontract interface development to a group specializing in this area could reduce or eliminate this as a risk.

It is important that the likelihood of risk occurrence and its impact be monitored throughout a project. The likelihood and impact of the risk can change over time. For example, the risk of losing a team member (due to health, resignation, or other causes) will typically have a bigger impact past the beginning of the project; but, depending on the team member's role, may have little affect at the end of the project. Also, monitoring should include identification of new risks. Both new risks and changes in risk exposure for old risks require re-planning on how to manage these changes.

EXERCISE 4.2: ASSESSING DH RISKS

After completion of the DH Task Plan and Schedule, Massood lead the team in a discussion session to support the development of a Risk Management Plan.

The exercise, described at the end of this chapter, deals with the DH Team's efforts in risk identification, analysis, and management.

Software Configuration Management

As Massood Zewail was wrapping up the DH project planning meeting in September 202X, Disha Chandra asked Massood "What about Configuration Management (CM)? Aren't we going to plan for that?" Massood said, "Oops, I thought that would be done separately?" Disha said, "No, I think we should do that as part of our initial planning activities". Massood said, "Sorry but I really don't know much about CM?" Georgia spoke up

> I took a training course in Software Configuration Management (SCM) when I worked at Volcanic Power, and I coordinated SCM on a couple of projects. If you like, I could lead a discussion on SCM and what we should do to plan for it.

Disha said, "Great, SCM is yours. Tell us what you know and lead the way".

MINITUTORIAL 4.5: SOFTWARE CONFIGURATION MANAGEMENT

During the development of a software system, even a modest sized one, things change: plans are altered; requirements are added or modified; changes are made to correct defects found in the design or in code; and incremental development results in different versions of the software artifacts. These changes need to be made in an orderly and organized fashion; otherwise, problems can arise: a unit developer may make changes to a unit's functionality without informing the rest of the development team, resulting in out of date requirements or design documents; a component developed from an earlier design version can be integrated into the current version of the system, producing a system without the required/expected capability; or defects discovered in system test might not be properly reported, causing the system to fail when put into operation.

Configuration management (CM) is concerned with managing the key items in the development, delivery, and maintenance of a computing system so that problems incurred because of the organization and structure of the system, changes to the system, and the interaction between system stakeholders do not unduly affect cost, quality, and schedule. The term *Software Configuration Management* (SCM) is used when we are dealing with a software system or subsystem. Table 4.4 describes some of the commonly used terms and abbreviations used in configuration management activities [IEEE 24765, Sommerville 2011, Bourque 2014].

Table 4.4 A Configuration Management Glossary

Term	Description
Baseline	Configuration information formally designated at a specific time during a product's or product component's life.
Configuration	The functional and physical characteristics of hardware or software as set forth in technical documentation or achieved in a product.
Configuration Control Board (CCB)	A group of people responsible for evaluating and approving or disapproving proposed changes to configuration items, and for ensuring implementation of approved changes. CCB sometimes is used to mean Change Control Board.
Configuration Item	An item that is designed to be managed as a single entity, within a configuration management system. (e.g., a project plan, a software requirements specification, source code for an object class, or a test plan).
Configuration Management (CM)	A discipline applying technical and administrative direction and surveillance to: identify and document the functional and physical characteristics of a configuration, control changes to those characteristics, record and report change processing and implementation status, and verify compliance with specified requirements. When the CM is concerned strictly with a software system or subsystem it is called SCM.

(Continued)

Table 4.4 (*Continued*) A Configuration Management Glossary

Term	Description
Release	A collection of configuration items which have been tested and are to be introduced (together) into a live operating environment. The release would typically include software documentation and executable code.
Software Building	The activity of combining the correct versions of software configuration items, by using the appropriate configuration data, into an executable program for delivery to a customer or other recipient, such as the testing activity.
Variant	A version of a configuration item not different in substance from the version it was derived from but providing a variation in the operational or application mode. For example, a variant of a version that runs on a different operating system.
Version	An initial release of a configuration item or a re-release of a configuration item that differs from previous versions of the item in terms of capability, structure, or requirements.

THE SCM PROCESS

The process used for SCM depends on a number of factors: the size and nature of the developing organization; the scope, complexity, and application domain of the software system; and whether the software will be developed as a stand-alone system or as part of a subsystem of a larger system. If the software is large and complex, involving many developers and a lengthy schedule, a robust well-defined process will be needed, like the plan-driven SCM process described in [Bourque 2014]. However, a smaller more self-contained project, with modest scope and schedule, would best be served by a simpler agile SCM approach.

A comprehensive SWEBOK SCM process would include the elements depicted in Figure 4.6:

- *Planning* – development and implementation of an SCM Plan.
- *Configuration Identification* – includes identifying the configuration items for a software development project.
- *Control Management* – controls changes to a configuration in an organized, disciplined manner.

■ *Status Accounting and Auditing* – status accounting concerns the recording and reporting of information needed for effective management of the software configuration; a software configuration audit is an examination by an independent body to determine the extent to which an item satisfies its functional and physical requirements
■ *Release Processing* – encompasses the identification, packaging, and delivery of the elements of a product – for example: executable program, documentation, release notes, and configuration data.

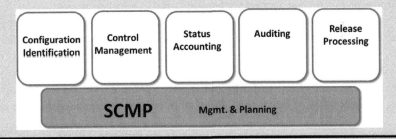

Figure 4.6   SCM process.

CONFIGURATION IDENTIFICATION

Software configuration identification identifies software system items to be controlled, establishes identification naming conventions for the items and their versions, and determines the methods and tools to be used in obtaining, storing, and managing controlled items. These activities provide the basis for the other SCM activities.

More specifically configuration identification includes the following activities:

■ Understanding the software configuration within the context of the system configuration
■ Choosing the software configuration items
■ Adopting a labeling strategy for software items
■ Creating and identifying the composition of baselines

The SCM process typically controls many items in addition to just the code. Possible configuration items include schedules and task plans, requirements and design specifications, testing materials, source and executable code, data and data dictionaries, and documentation for installation, maintenance, operations, and software use. The decision about what to select as configuration items depends on the scope and complexity of the project and the resources needed to manage a configuration. For example, looking at the Requirements Engineering part of the TMS Gantt Chart in Figure 4.3, the TMS project team would have to decide which analysis documents (use case diagrams, sequence diagrams, class diagrams, etc.) would be placed under SCM; or it may be decided to combine all analysis documents as a single configuration item.

A software baseline represents the set of the established versions of software configuration items at a specific point during a software development life cycle. Software configuration items are placed under SCM control at different times and are incorporated into a particular baseline at a particular point in the software life cycle. The triggering event is typically the completion of some type of a formal acceptance activity, such as a formal review.

Figure 4.7 characterizes the evolution as a life cycle proceeds. Notice that the changes in the baseline seem to follow a waterfall life cycle. An incremental or agile process might follow a less ordered and more flexible progression. Orderly and accurate changes in a baseline are dependent on controlled software libraries and effective SCM tool support, especially with processes involving frequent software builds (e.g., daily builds or multiple builds within a day).

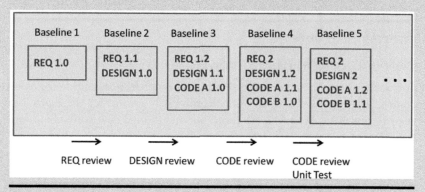

Figure 4.7   Baseline evolution.

CONTROL MANAGEMENT

Software configuration control is concerned with managing changes during the software life cycle. It covers the process for determining what changes to make, the analysis and approval of configuration changes, managing the implementation of those changes, and verification and acceptance of the changes. Measurement of these activities is useful in evaluating the effectiveness of the SCM process, assessing the cost of rework, and determining how the SCM and life cycle processes can be improved.

Figure 4.8 (derived from [Bourque 2014]) depicts a process that could be used to manage a change in a configuration item. The process begins with the generation of a Software Change Request (SCR), based on a perceived need for change (e.g., a customer request for a change in requirements, the discovery of a design defect in a design review, or a code defect is found from a unit test). The SCR is reviewed by a CCB and, if approved, is implemented and verified. For small projects, the CCB might consist of a single designated individual. However, even for small projects some process for managing change should be instituted to ensure that changes are properly communicated, implemented, and verified. Lack of an effective change process can result in cost, quality, and schedule problems. Well-defined SCM is especially critical for distributed and global projects, in which communication and change management can be challenging because of geographical, time zone, and cultural differences; such differences add complexity to SCM and can lead to confusion and error if not properly managed.

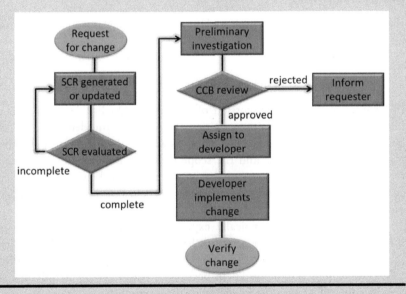

Figure 4.8 Change control process.

EXERCISE 4.3: DETERMINING THE CONFIGURATION IDENTIFICATION

After Georgia gave a brief overview of the SCM process and discussed the typical SCM methods and activities, she asked the team to consider what software configuration items should be included in the DH configuration.

The exercise, described at the end of this chapter, deals with the DH Team's efforts to determine what the configuration of the DH should look like: that is, to identify the DH configuration items.

EXERCISE 4.4: DEVELOPING A SCM CHANGE PROCESS

After the DH Team completed the configuration identification activity, Georgia suggested that they next work on a SCM change process for the project. Yao Wang spoke up and said he did not see the need for a formal change process for such a small project. However, Michel Jackson disagreed – he said he had seen too many projects, both large and small, get into serious problems because they did not manage change properly. The team decided to develop a change process.

The exercise, described at the end of this chapter, deals with the DH Team's efforts to develop an SCM change process.

Quality Planning

As the final activity in the DH project planning meeting, Massood led a discussion about quality planning.

MINITUTORIAL 4.6: QUALITY PLANNING

In Chapter 3, there was discussion of a Software Quality Assurance Plan (SQAP). During the planning activities, the project team identifies a set of quality goals and develops a plan for how to achieve those quality goals. The plan includes the purpose and the scope of the quality activities, and the required resources and processes that are needed to achieve those quality goals. In addition, roles and responsibilities of the people involved in quality assurance activities are defined.

The quality goals are typically expressed in terms of metric (perhaps through GQM approach) benchmark levels to be achieved by the quality activities. For example, the SQAP for TMS project might have the quality activity of conducting a formal inspection of the Software Requirements Specification (SRS) with the following quality goals:

- a defect removal rate of 0.5 defects/hour or higher;
- a review rate of 2 pages/hour or less;
- and a yield rate (percent of total defects removed) of 70% or higher.

EXERCISE 4.5: DEVELOPING A QUALITY PLAN

After the DH Team completed the planning for configuration management, they proceeded to develop a quality plan.

The exercise, described at the end of this chapter, deals with the DH Team's efforts to develop a quality plan.

Case Study Exercises

EXERCISE 4.1: MAKING A TASK PLAN AND SCHEDULE

In late September 202X, the DH Team held a project planning meeting in order to develop a WBS, and a task plan and schedule.

Learning Objectives

Upon completion of this exercise, students will have increased ability to:

- Estimate the size of (in LOC) and the required total effort (in hours) of a moderate-sized software product.
- Determine the WBS for a software development effort – i.e., a description and organization of the software engineering tasks that need to be completed to develop a software product.
- Estimate the time needed to complete various software engineering tasks.
- Develop a project task plan and schedule for development of a software product.

DH Case Study Artifacts

- DH High-Level Requirements Definition (HLRD)
- DH Development Process Script
- DH Conceptual Design

Exercise Description

1. As preparation for the exercise, read Chapter 1 and the Case Study Artifacts listed above.
2. You will be assigned to a small development team (like the DH Team).
3. Each team meets, organizes, and carries out the following activities:
   a. Develop a WBS for the DH System (use the HLRD and the Conceptual Design to determine the components).
   b. Estimate the size (in LOC) of the DH product.
      - Use the Size-Effort Estimating Template in the DH Planning Tool
      - List each component to be developed (use the WBS).
      - For each component listed in the template, estimate and enter the number of LOC to be developed.
      - The total estimated size will be computed by the tool.
      - Note: An alternate approach would be for each team member to make an independent estimate and use a Delphi Wideband technique.
   c. Estimate total effort (in hours) required to complete the project.
      - Use the Size-Effort Estimating Template in the DH Planning Tool
      - Estimate productivity of a team member in LOC/hour. (Industrial data shows an engineer, on average, develops 5 – 10 LOC/day.)

- Divide the estimated LOC for each component by LOC/hour estimate and enter the value in the effort column of the Size-Effort Estimating Template.
- The total estimated effort will be computed by the tool.

d. Next, complete the Task-Schedule Planning Template of the DH Planning Tool.

- List the task phases (extracted from DH Development Process Script) and tasks within each phase that need to be completed. Begin your plan with the Analysis phase of the DH development.
- Note: The DH Development Process should be helpful. In some cases, you can copy and paste tasks from the process table to the Planning Template.
- For each task, specify the following:
  - A team member as lead for the task
  - An estimated date of completion for the task (between late September 202X and July 201Y)
  - The estimated required effort to complete the task (in hours)
  - The PV for the task

4. Choose one member of the team to report to the class on the team's planning activity.

EXERCISE 4.2: DEVELOPING A RISK MANAGEMENT PLAN

In late September 202X, the DH Team held a project planning meeting; part of the team's activities included developing a risk management plan: identifying risks, analyzing them, and planning on how to manage the risks

LEARNING OBJECTIVES

Upon completion of this exercise, students will have increased ability to:

- Identify risks for a moderate-sized software development project.
- Analyze risks by estimating risk likelihood and risk impact.
- Use risk exposure to prioritize risks.
- Plan ways to manage risks.

DH CASE STUDY ARTIFACTS

- DH Development Process Script
- DH High Level Requirements Definition
- DH Conceptual Design

EXERCISE DESCRIPTION

1. As preparation for the exercise, read Chapter 1 and the Case Study Artifacts listed above.
2. You will be assigned to a small development team (like the DH Team).

3. Each team meets, organizes, and carries out the following activities:
   a. Perform a risk analysis for the DH project (use the DH Risk Planning Template):
      • Identity the risks (including security risks) and describe them.
      • Estimate the occurrence likelihood for each risk – from 1 (very unlikely) to 5 (very likely).
      • Estimate the impact if the risk occurs, for each risk – from 1 (very low impact) to 5 (very high impact).
      • Compute the Rick Exposure
      • Order the risks from highest risk exposure to the lowest
   b. Make a plan for managing the risks identified:
      • For each risk, determine how it should be managed (avoidance, control, assumption, transfer).
      • Enter the management decision in the DH Risk Planning Template.
4. Choose one member of the team to report to the class on the team's risk planning activity.

EXERCISE 4.3: DETERMINING THE CONFIGURATION IDENTIFICATION

In late September 202X, the DH Team held a project planning meeting; part of the team's activities included planning for SCM. There first task was to carry out configuration identification for the DH project.

LEARNING OBJECTIVES

Upon completion of this exercise, students will have increased ability to:

• Understand the nature and organization of a software configuration that would be placed under management.
• Identify the software configuration items for a moderate-sized software development project.
• Design a naming and labeling system for software configuration items.

DH CASE STUDY ARTIFACTS

• DH Development Process Script
• DH High Level Requirements Definition
• DH Conceptual Design

EXERCISE DESCRIPTION

1. As preparation for the exercise, read Chapter 1 and the Case Study Artifacts listed above.
2. You will be assigned to a small development team (like the DH Team).
3. Each team meets, organizes, and carries out the following activities:
   a. Identify the software configuration items to be used in a DH SCM.

    b. Decide how the configuration items will be named and how different versions of the same item will be labeled.

4. Choose one member of the team to report to the class on the team's configuration identification decisions.

EXERCISE 4.4: DEVELOPING A SCM CHANGE PROCESS

In late September 202X, the DH Team held a project planning meeting; part of the team's activities included planning for SCM. One of the tasks was to develop a SCM change process for the DH project.

LEARNING OBJECTIVES

Upon completion of this exercise, students will have increased ability to:

- Understand the nature and activities associated with managing a change in a software configuration item.
- Analyze the advantages and disadvantages of a SCM change process for a given project.
- Design a change management process for a given project.

DH CASE STUDY ARTIFACTS

- DH Development Process Script
- DH High Level Requirements Definition
- DH Conceptual Design

EXERCISE DESCRIPTION

1. As preparation for the exercise, read Chapter 1 and the Case Study Artifacts listed above.
2. You will be assigned to a small development team (like the DH Team).
3. Each team meets, organizes, and carries out the following activities:
   a. Discusses what activities and details should be part of the DH SCM change management process.
   b. Describe the details of the DH change management process: process steps, forms and tools to be used, the makeup of the CCB, how implemented changes will be verified, and what records/data will be retained.
4. Choose one member of the team to report to the class on the team's SCM Change Process.

EXERCISE 4.5: DEVELOPING A QUALITY PLAN

In late September 202X, the DH Team held a project planning meeting; part of the team's activities included planning for SCM. One of the tasks was to develop a quality plan for the DH project.

Learning Objectives

Upon completion of this exercise, students will have increased ability to:

- Understand the nature and activities associated with developing a quality plan for a moderate-sized software development project.
- Decide on the quality activities for a moderate-sized software development project.
- Determine the quality goals (using benchmark metrics) for the software quality activities for a moderate-sized software development project.

DH Case Study Artifacts

- DH Development Process Script
- DH High Level Requirements Definition
- DH Conceptual Design

Exercise Description

1. As preparation for the exercise, read Chapter 1 and the Case Study Artifacts listed above.
2. You will be assigned to a small development team (like the DH Team).
3. Each team meets, organizes, and carries out the following activities:
   a. Decides on what quality activities will be carried out for the DH project (e.g., an SRS inspection).
   b. For each quality activity selected, determines the quality goal(s) (i.e., benchmark metric levels, such as defect yield for the activity).
4. Choose one member of the team to report to the class on the team's quality planning activity.

Chapter 5

Engineering the DH Requirements

Software Requirements Fundamentals

In early October of 202X, Jose Ortiz and Disha Chandra met to review the progress of the DH Project Team. The Team had completed their Project Launch (discussed in Chapter 2), and the initial planning activities were nearing completion (discussed in Chapter 4). They both thought the project was progressing nicely: they were pleased with the development of the *DH Customer Need Statement*, the *High-Level Requirements Definition* (HLRD), and the *DH Development Process*; they felt the *DH Conceptual Design* and the *DH Development Strategy* provided excellent support for project planning; and the project team was beginning to "jell", establishing effective working relationships.

The team was ready to begin the Analysis Phase of the project. Disha and Jose reviewed the tasks to be performed in this phase, listed in the *DH Development Process* (and reproduced in Table 5.1). Jose confessed to Disha that he had some anxiety about the upcoming phase. He had seen too many projects that floundered or failed because of inadequate attention to the importance of requirements. Disha suggested that since Michel Jackson would be leading the DH Analysis phase and had vast experience in requirements engineering that he be asked to start the Analysis phase with a tutorial session on the development of software requirements.

Disha conferred with Michel about the requirements tutorial session; Michel thought it was a good idea and said he would get started on developing the tutorial right away. The following is a summary of what Michel covered in his requirements tutorial session with the team.

Table 5.1   Analysis Phase Tasks

1. Review the Need Statement and prepare for requirements elicitation.
2. Elicit requirements from potential customers and users.
3. Analyze requirements and build object-oriented analysis models.
4. Specify and document requirements in an SRS.
5. Inspect and revise the SRS.
6. Obtain customer agreement to requirements.
7. Develop a System Test Plan.
8. Develop an RTM (Requirements Traceability Matrix.)
9. Develop a User Manual.
10. Track, monitor, and update plans (Task/Schedule/Quality/CM/Risk) & RTM.

**MINITUTORIAL 5.1: GETTING STARTED
WITH REQUIREMENTS**

Agreement on software requirements (what the software is supposed to do) is one of the most critical parts of software development. As Fred Brooks stated: "The hardest single part of building a system is deciding what to build… No other part of the work so cripples the resulting system if done wrong. No other part is more difficult to rectify later" [Brooks 1995].

DEFINITIONS

- The IEEE Standard Systems and Software Engineering Vocabulary [IEEE 24765] defines a *software requirement* as:
  - A software capability needed by a user to solve a problem to achieve an objective.
  - A software capability that must be met or possessed by a system or system component to satisfy a contract, standard, specification, or other formally imposed document.

System requirements are the requirements for the system as a whole. For example, a Traffic Management System (TMS) would have requirements for the various system elements:

- people (e.g., requirements for the traffic police that help control traffic in the TMS)
- hardware (e.g., requirements for the signage and traffic signals in the TMS)
- software (e.g., requirements for the software subsystem of the TMS that helps manage and control traffic)

One common way to classify requirements is as "functional" or "non-functional". From SWEBOK 2014 [Bourque 2014]:

- *Functional requirements* describe the functions that the software must execute. (e.g., "The TMS shall provide the capability for a Traffic Manager to set the duration of individual traffic signal states using the following duration ranges: Red [30–120 seconds], Green [30–120 seconds], and Amber [5–10 seconds]".)
- *Non-functional requirements* are the ones that act to constrain the solution. (e.g., "The software for the TMS shall be written in Java" or "The TMS software subsystem shall not fail more often than once in every 24,000 hours".)

A *Quality Attribute* is a characteristic that is related to the performance or use of the software, which is not related to specific functionality of the software. Example attributes include the following: availability, reliability, usability, reusability, maintainability, efficiency (in use of computing resources), safety, and security.

An *Emergent Requirement* is a requirement which cannot be addressed by a single component, but which depends for its satisfaction on how all the software components interoperate [Bourque 2014] (e.g., "The TMS shall provide for at least a 20% increase in traffic throughput during the weekday traffic period of 4:30 pm to 6:30 pm".).

To better specify the quality attributes of the DH system, Michel suggested the team use *quality attributes scenarios* [Clements 2010] as a method for documenting a quantifiable and analyzable set of non-functional requirements. Since most of the DH Team was not familiar with quality attributes scenarios, Michel prepared and delivered a short tutorial on the subject.

MINITUTORIAL 5.2: QUALITY ATTRIBUTE SCENARIOS

It is the case that architecture and design are most often concerned with moving from functional requirements to a developed system (i.e., given a set of inputs it produces the correct answer or responds with the correct functionality). However, software architecture must be concerned with more than just functionality. It must address competing and conflicting influences. Such influences are typically the non-functional requirements (quality attributes). Even though these quality attributes are as important as functional requirements, they may not be formally specified. Quality attributes include attributes related to system performance, availability, security, reliability, modifiability, testability, and usability.

The purpose of a software quality attribute scenario is to provide a way of identifying and documenting the quality attribute in an unambiguous and quantifiable manner. As mentioned in MiniTutorial 5.1, the architecture must be described and documented in a way that allows for the analysis of quality attributes and whether the architecture meets these quality attributes. Such analysis is impossible without clearly defined quality attributes.

In this tutorial, we use quality attribute scenarios to specify a system quality attribute. A scenario consists of the following six fields:

1. *Stimulus* – the actual event that the system must respond to
2. *Source of Stimulus* – the entity responsible for generating the stimulus
3. *Environment* – the conditions under which the stimulus occurs (different environments might result in different system responses)
4. *Artifact* – the part of the system being stimulated. This could be the whole system, or a single artifact (e.g., source code, a test plan, ...)
5. *Response* – the desired activity that must be performed, because of the stimulus
6. *Response Measure* – a way of making the response quantifiable

[Bass 2012] provides descriptions of scenarios for the performance, availability, security, modifiability, testability, and usability. It also provides details on the typical values for each of the six fields that make up a scenario for each quality attribute. Below, we provide an example availability scenario for the DH system. Availability quality attributes are concerned with the reaction of the system to failures (e.g., what parts of the system will be unavailable and for how long).

EXAMPLE RELIABILITY QUALITY ATTRIBUTE SCENARIO

The *DigitalHome Software Requirements Specification*, Version 1.2, in Appendix B, has the following Reliability requirement: "The *DigitalHome* System must be highly reliable with no more than 1 failure per 10,000 operations".

BREAKDOWN OF THE SCENARIO

- *Source* – users, Internet service, DH web server, DH gateway device, DH sensors, DH controllers, DH software, physical environment
- *Stimulus* – device failure, interruption of Internet service, software fault, security violation
- *Environment* – system setup, normal operation, over-loaded operation, system maintenance
- *Artifacts* – gateway processor, communication channels, failed device, system database, security information
- *Response* – software exception raised, system goes from normal operation to maintenance mode, DH technician takes control of the system
- *Response Measure* – number of operations performed since the last failure

EXERCISE 5.1: SPECIFICATION OF A QUALITY ATTRIBUTE SCENARIO

After Michel had completed his presentation on Quality Attributes, he led the team in specifying a quality attribute scenario for security requirement for the *DigitalHome* system. The exercise, described at the end of this chapter, deals with how the Team developed the scenario.

REQUIREMENT CHARACTERISTICS

The development of correct and effective requirements is critical to the success of a software system. Problems with software requirements are one of the leading causes of schedule and budget overruns and can have significant impact on the quality of the delivered product or service [Standish 2020].

The following are characteristics of an effective set of software requirements [IEEE 830]:

- *Correct* – every specified requirement is one that the software must satisfy
- *Clear/Unambiguous* – requirements are expressed in a precise, unambiguous, easy to understand format; there is only one interpretation of a requirement
- *Complete* – includes all requirements agreed upon by the system stakeholders
- *Concise* – describes a single property, without excessive verbiage
- *Consistent* – no two requirements contradict each other
- *Feasible* – realizable within resource and schedule constraints
- *Traceable* – can be traced backward to sources (e.g. end user, customer, government, or company policies) and forward to software artifacts developed from the requirements (e.g., design components, code, test cases)
- *Verifiable* – has a clear, testable criterion, and a cost-effective way to check that a requirement has been properly implemented
- *Prioritized* – ranked for importance and/or stability (e.g., for importance: "essential", "conditional", "optional")

THE REQUIREMENTS PROCESS

The basic goal of the software requirements process is to collect, analyze, specify, and validate what the software is to do, without describing how the software will implement the requirements. Such processes vary in detail and formality depending on the scope, size, and domain of the software product or service being developed.

Figure 5.1 depicts four subprocesses that would be part of most software requirements processes:

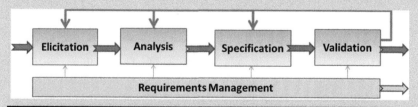

Figure 5.1   Requirements process.

- *Requirements Elicitation* – determine where software requirements come from (identifying stake holders) and collect them.
- *Requirements Analysis* – study, analyze, and model the problem to be solved

Note: The Analysis Phase for the DH Project (described in Table 5.1) includes "analyze requirements and build object-oriented analysis models". Hence, the DH Analysis Phase includes the above "Requirements Analysis" subprocess.

- *Requirements Specification* – define and document the requirements in an organized and precise manner
- *Requirements Validation* – ensure that the requirements are properly understood and confirm that the requirements conform to applicable standards and regulations, and that they are understandable, correct, consistent, and complete.
- *Requirement Management* – for a complex software system, the above processes much be managed throughout the life of the software, because the requirements will change – new ones added and modifications made to improve understanding or correct errors.

Notice in Figure 5.1 that requirements development is not strictly linear: work in one phase of the process may lead to a return to an earlier phase. For example, during analysis of the TMS requirement on the duration of different stoplight states (red, green, amber), an analyst may question whether the periods are long enough for some situations. In this case, the engineer may need to go back and discuss this with traffic officials; hence, reentering the Elicitation Phase.

Also, notice in Figure 5.1 that the Requirements Management activities extend throughout the Requirements Process and beyond. In Chapter 4, the discussion of configuration management would apply in particular to the management of software requirements and their changes, and thus would be part of the Requirements Management process.

If a project is using an incremental process, it might repeat requirements activities for each increment. In fact, even with a waterfall-type approach, it is not uncommon for a development team to have to repeatedly return to the requirements process in a later phase of development –in design, construction, or testing.

Eliciting Requirements

One of the key decisions at the August 202X *HomeOwner* retreat was for *DigitalHomeOwner*, with the assistance of the Marketing Division, to conduct a needs assessment for a *DigitalHome* product, producing the *DH Customer Need Statement* and the HLRD. These efforts might be considered the beginning of the DH requirements elicitation. However, the primary purpose of the documents produced was to provide guidance for launching the DH project and to supply enough information about the DH nature and scope to support effective planning activities (Chapter 4). The DH Team felt the Need Statement and HLRD lacked the detail and certainty needed to adequately specify the DH requirements.

After Michel's requirements tutorial, Jose, Disha, and Michel met and discussed the need for more specificity about which features of the DH should be included. The decision was made for the DH Team to carry out a formal requirements elicitation effort led by Michel.

MINITUTORIAL 5.3: ELICITATION TECHNIQUES

Eliciting software requirements is an essential part of determining "what" the software is supposed to do, in addition to determining any development constraints or other non-functional requirements. There are a variety of techniques used to elicit requirements, including in the following:

INTERVIEWS

Interviews are probably the most traditional and commonly used technique for requirements elicitation. There are various types of interviews:

- *One-on-One* – one analyst (or more) interviews one customer or user.
- *Facilitated Meeting/Workshop* – discussion about requirements between developers, and users and other appropriate stakeholders of a system, led by a facilitator
- *Un-facilitated Meeting (Brainstorming)* – an informal version of workshop, which allows stakeholders to propose any requirement, thereby allowing the discovery of hidden requirements.
- *Closed/Structured Interviews* – the requirements analyst solicits answers to a pre-defined set of questions.
- *Open/Unstructured Interviews* – there is no predefined agenda and the requirements analyst discusses, in an open-ended way, what stakeholders want from the system.

APPRENTICING (ROLE PLAYING)

This technique involves the requirements analyst actually learning and performing the tasks of the user (which are related to the software being developed) under the supervision of an experienced user. Apprenticing is very useful where the analyst is inexperienced with the domain and when the users have difficulty in explaining their actions. The apprentice can serve as surrogate user throughout the rest of development process.

OBSERVATION (SHADOWING)

Analysts learn about user tasks by immersing themselves in the environment and observing how users interact with the software and with each other (Figure 5.2). It is good for discovering patterns of use between parts of an organization.

Observation is similar to apprenticing, but without direct involvement or inter-ference with user tasks and business processes. The effectiveness of observation can vary as users tend to adjust the way they perform tasks when knowingly being watched.

USE CASES AND STORYBOARDING

A *Use Case* (UC) is a structured description of some part of the functionality of a software system (see Chapter 3). Use cases are used to communicate with a user, eliciting requirements or checking on whether the required functional-ity has been properly captured.

Storyboarding, the development of storyboards, is a more informal tech-nique used to elicit user/customer requirements. A software *storyboard* is a graphic in the form of screens that the software will display; they are drawn, either on paper or using specialized software, to illustrate the important steps of the user experience. The storyboard is then modified by the engineers and the client while they decide on their specific needs.

PROTOTYPING

A *prototype* is an initial simplified version of a system which may be used for experimentation. Prototypes are valuable for requirements elicitation because users can interact with the system and point out its strengths and weaknesses. They provide something concrete to evaluate. A simple graphic interface proto-type is similar to a storyboard.

A prototype can be used to establish feasibility and usefulness before high development costs are incurred, and can be helpful in developing the "look and feel" of a user interface. There are a wide range of prototyping techniques from paper mock-ups of screen designs to beta-test versions of software prod-ucts. There is a strong overlap of their use for requirements elicitation and the use of prototypes for requirements validation.

Figure 5.2   Eliciting requirements by observation.

EXERCISE 5.2: ELICITING DH REQUIREMENTS

After it was decided that the DH Team would carry out a formal elicitation effort, Michel's first action was to suggest a team meeting devoted to planning for the requirements elicitation effort. The exercise, described at the end of this chapter, deals with how the DH Team will plan for its elicitation activities.

Analyzing Requirements

After the DH Team completed its initial requirements elicitation activities, using the elicitation plan it developed, Michel Jackson led a discussion of the next task listed in the *DH Development Process*: "analyze requirements and build object oriented analysis models". Michel made the following points:

- The *purpose of requirements analysis* is to take information provided by the stakeholders, and analyze and model that information so that
  - we better understand the DH stakeholder needs and requirements;
  - the DH functional and non-functional requirements can be specified clearly, correctly, and completely;
  - and, there is a solid foundation for our design, construction, and testing of the DH software.
- During the requirements analysis phase, we will build a "conceptual model" for our system that will support effective communication between potential *DigitalHome* users, *HomeOwner* management, and our team. For our purposes, I recommend that the DH Conceptual Model consists of the following elements:
  - *Context Diagram* – a view of system boundaries, external entities that interact with the system and the relevant information flow between these entities and the system.
  - *Conceptual Design* – an initial view of the DH system architecture with a high-level view of the principal components and their relationship.
  - *Use Case Model* – a model that describes the various ways (use cases) that system users can interact with a system.

Yao Wang, the DH System Architect, suggested that the team consider adding some additional features to the DH Conceptual Model by expanding some of the Conceptual Design components into class diagrams and adding sequence diagrams corresponding to the use case scenarios. (See Chapter 6 for information on class and sequence diagrams.)

Michel responded that he thought the model that he had proposed would be sufficient:

■ since this was a prototype version, the model recommended would provide the necessary analysis; in his experience, the use case model had proved to be an excellent analysis tool for the size and scope of a project like *DigitalHome*;

■ and the limited development time for *DigitalHome* would not allow for a full-scale, expansive analysis; and since they had already completed a Context Diagram and Conceptual Design (Chapter 2), this would allow adequate time to develop a high-quality Use Case Model.

After some discussion, the DH Team agreed to follow Michel's recommendation and proceeded with developing the *DH Use Case Model* (available in Appendix C).

MINITUTORIAL 5.4: USE CASE MODELING

A *Use Case* (UC) is a structured description of some part of the functionality of a software system. It describes a way in which the system is used.

A use case *actor* is a user or some other external entity (e.g., another system such as a database or an alarm system) that interacts with the system being analyzed. A use case includes a *scenario*, which provides a sequence of interactions between the system being modeled and its external actors.

Use cases are most helpful in requirements elicitation and analysis. Since the use case scenario provides a picture of how the user interacts with the system, an initial version can be used to elicit user response that will help clarify, expand on, or correct the system functionality being modeled. In its detailed mature form, a Use Case Model can be used to specify the functional requirements. Use cases can also be used in system test planning: use case scenarios can be converted into test case scenarios that guide the engineer performing the test.

There are different opinions, approaches, techniques, and styles associated with use case modeling. Although use case modeling is part of the Unified Modeling Language (UML) (discussed in Chapter 6) and UML supports object-oriented modeling, a Use Case Model has a functional character and could be used to support a non-object-oriented functional modeling approach.

UML includes a diagram, a *Use Case Diagram*, for depicting actors, uses case, and their relationships [Fowler 2004]. Figure 5.3 shows a use case diagram for the TMS, described in Chapter 3. The stick figures represent the actors, the ovals represent the use cases, and the connectors represent the relationships between the actors and the use cases. UML provides for different types of relationships between use cases (*include*, *generalization*, and *extend*) [Fowler 2004]. Notice the actors include non-human entities that interact with the system (e.g., the traffic signals). Also, note that relationships with the Traffic Signals and Traffic Cameras are labeled "1..*"; this "multiplicity" label indicate "one or more" traffic devices are part of the TMS.

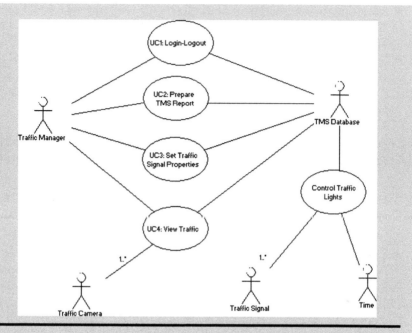

Figure 5.3   TMS use case diagram.

A Use Case Model includes a *scenario* for each use case: a sequence of steps describing the interaction between the actor(s) and the system. There is no standard for writing use case scenarios. Some are less structured, just listing the scenarios steps; and others have different or additional components. The DH Team used a Scenario with the following features:

- a use case **Goal**
- a list of **Actors**
- a pre-condition (**Pre**), which states what must be true before carrying out the use case scenario.
- a post-condition (**Post**), which states what must be true when the use case scenario is completed.
- a **Main Success Scenario** that describes a set of steps which will lead to achievement of the uses case Goal. The steps are numbered and may include control structures such as selection (e.g., steps 13A and 13B) and repetition (e.g., Go to Step 1).
- **UC GUIs** that are referenced in the scenario.
- **Exceptions (Alternative Scenarios of Failure)** that describe what happens when an exceptional situation occurs, which would interrupt the Main Success Scenario and require an alternate scenario. In such a situation, the use case Goal may not be achieved.

- **Use Cases Utilized** in the scenario. A scenario can call (or "include") another use case, which is equivalent to inserting the referenced use case scenario steps into the main scenario.

Figure 5.4 shows a use case scenario for the use case *UC3: Set Traffic Signal Properties*.

Use Case ID: UC3
Use Case Name: Set Traffic Signal Properties
Goal: Set/Change traffic signal properties
Primary Actors: Traffic Manager
Secondary Actors: TMS Database
Pre:
- Traffic Manager is logged into his/her account.
- TMS Traffic Signal Configuration Page is displayed on User Display Device.

Post:
- All newly entered traffic signal properties have been stored in the TMS Database.
- TMS Traffic Signal Configuration Page is displayed on User Display Device.

Main Success Scenario:

Step	Actor Action	Step	System Reaction
1	Select "Set Traffic Signal Properties"	2	Display "Select the traffic signal(s) for which you wish to set properties".
3	Select one or more traffic signals.	4	Display "Enter Green Light duration (in seconds)".
5	Enter seconds of Green Light duration.	6	Store Green Light duration for selected traffic signals in the TMS database.
		7	Display "Enter Red Light duration (in seconds)".
8	Enter seconds of Red Light duration.	9	Store Red Light duration for selected traffic signals in the TMS database.
		10	Display "Enter Amber Light duration (in seconds)".
11	Enter seconds of Amber Light duration.	10	Store Amber Light duration for selected traffic lights in the TMS database.
		12	Display "Do you wish to set the properties for other traffic signals? YES/NO?"
13A	Enter "YES"	14A	Go to Step 1
13B	Enter "NO"	14B	Display TMS Traffic Signal Configuration Page

UC GUIs: DH Configuration Page, DH Default Parameter Configuration Page
Exceptions (Alternative Scenarios of Failure):
- At any time, User fails to make an entry.
  a. Timeout after 5 minutes and logout from the system.
- System detects a failure (loss of internet connection, power loss, hardware failure, etc.).
  a. Display an error message on the User Display Device.
  b. System restores system data from TMS Database.
  c. Logoff user.

Use Cases Utilized: none
Notes and Issues: The duration of individual traffic signal states must be in the following duration ranges: Red [30 to 120 sec], Green [30 to 120 sec], and Amber [5 to 10 sec]

Figure 5.4 TMS use case scenario.

EXERCISE 5.3: MODELING A DH USE CASE

In November of 202X, Red Sharpson, the CEO of Homeowner, was visiting one of his stores, walking the aisles, talking to customers. While engaged in a conversation with one customer, he mentioned Homeowner's *DigitalHome* (DH) project. The customer asked Red if they were going to have motion sensors that could automatically turn lights on and off. Red had recently read a draft of the *DigitalHome Software Requirements Specification* and replied that they currently had no such plans. That night at dinner he told his wife Wilma about the customer's idea. Wilma loved the idea – "Red, you absolutely have to include this feature. It is a clear winner!" Afterwards, Red contacted Dick Punch, the *HomeOwner* VP of Marketing, asked him to conduct a study to see if motion sensors would be an attractive feature for *DigitalHome*.

Dick had a quick telephone survey conducted and found that there was great interest in a motion sensor feature. After this was reported to Red, he directed Jose Ortiz to add a motion sensor capability to the development of the *DigitalHome* prototype.

Subsequently, Jose asked Disha Chandra, the DH Team Leader, to conduct an analysis of the motion sensor feature and revise the Use Case Model, adding the feature.

The exercise, described at the end of this chapter, deals with the Team's analysis of the motion sensor feature, including use case modeling of the feature.

Specifying Requirements

As the DH Team neared completion of the DH Use Case Model, they began discussing how they would specify the DH Requirements. Yao Wang, the DH System Architect, felt the Use Case Model would be sufficient – he asserted "it provides sufficient detail to create a design and since this is a prototype project, developing full-scale requirements specification would not be necessary". However, Disha pointed out that the DH initial charge included the directive: *It is antici-pated that the DigitalHome prototype will be the foundation for future development of DigitalHomeOwner products. Hence, it is essential that the development team use established software development practices and fully document their work.* She stated it was necessary that the team develop a high-quality, comprehensive Software Requirements Specification (SRS).

After this, Michel led a discussion of the nature and format for the SRS. The team reviewed the IEEE Standard on *Recommended Practice for Software Requirements Specification* [IEEE 830] and other sources [Sommerville 2011, Wiegers 2003] to determine the SRS organization. The team agreed on the format in Figure 5.5, and over the next week produced the *DigitalHome Software Requirements Specification*, Version 1.0. Subsequently, minor changes were made and Versions 1.1 and 1.2 fol-lowed. The DH Configuration Management system (Chapter 4) was used to man-age changes to the SRS.

1. Introduction
2. Team Project Information
3. Overall description
   3.1 Product Description and Scope
   3.2 Users Description
   3.3 Development Constraints
   3.4 Operational Environment
4. Functional Requirements
   4.1 General Requirements
   4.2 Thermostat Requirements
   4.3 Humidistat Requirements
   4.4 Appliance Management Requirements
   4.5 DH Planner Requirements
5. Other Non-Functional Requirements
   5.1 Performance Requirements
   5.2 Reliability
   5.3 Safety Requirements
   5.4 Security Requirements
   5.5 Maintenance Requirements
   5.6 Business Rules
   5.7 User Documentation:
6. References
Appendix – Use Case Diagram

Figure 5.5   SRS outline.

MINITUTORIAL 5.5: SOFTWARE
REQUIREMENTS SPECIFICATION

PURPOSE AND VALUE

The IEEE Standard on *Software Life Cycle Processes* [IEEE 12207] states that "The *system requirements specification* shall describe: functions and capabilities of the system; business, organizational and user requirements; safety, security, human-factors engineering (ergonomics), interface, operations, and maintenance requirements; design constraints and qualification requirements. The system requirements specification shall be documented". For a software system, the systems requirements would consist of only software requirements; for more complex systems (software and hardware), the software requirements would be subset of the system requirements, and documented in an SRS.

The IEEE Standard on *Recommended Practice for Software Requirements Specification* [IEEE 830] describes recommended approaches for the specification of software requirements. A good SRS should provide several specific benefits:

- Establish the basis for agreement between the customers and the suppliers on what the software product is supposed to accomplish. The complete description of the functions to be performed by the software will assist the potential users in determining if the software specified meets their needs.
- Provide a baseline for validation and verification. Organizations can develop their validation and verification plans much more productively from a good SRS.
- Provide a basis for estimating costs and schedules. The description of the product as given in the SRS should provide a realistic basis for estimating project costs and can be used to obtain approval for bids or price estimates.

Agile processes, such as Scrum and XP, do not use a comprehensive requirements process, as described in [IEEE 12207]. Rather than use an [IEEE 830] approach to developing an SRS, Scrum requirements (called stories) are described in a "Sprint Backlog" prior to implementation, and then maintained in a "Product Backlog", as they are implemented. In XP, requirements are defined incrementally and expressed as automated acceptance tests rather than specification documents.

SRS ORGANIZATION

A recommended outline for the SRS, similar to Figure 5.5, is included in [IEEE 830]. The standard also provides other approaches (with a corresponding set of templates) for organizing software requirements:

- *System Mode* (e.g., the TMS, described in Chapter 3, might have different modes of operation – light traffic, normal, high traffic, or emergency)
- *User Class* (e.g., the TMS has different user types: traffic managers, drivers, pedestrians, etc.)
- *Objects* (e.g., the TMS objects include traffic signals, sensors, video cameras, roads, etc.)
- *Feature* (e.g., the TMS may have different features such as morning rush hour, normal operation, weekend traffic, etc.)
- *Stimulus* (e.g., the functions for the TMS might be organized into sections such as loss of power, emergency vehicle traffic, major sporting event, etc.)

REQUIREMENTS STYLE

Software Requirements can be written in a variety of styles and format, over a spectrum of formality. In the last 20 years, there has been a great deal of research and development devoted to how to structure and model software requirements. Table 5.2 describes some of these approaches.

GUIDELINES FOR WRITING REQUIREMENTS

MiniTutorial 5.1 on "Getting Started with Requirements", defined various types of requirements and stated some of the important attributes of an effective set of software requirements (clear, concise, correct, consistent, verifiable, etc.). So, the question naturally arises "How does one go about writing a good set of requirements?" There are many good sources offering advice and guidance about writing software requirements [IEEE 830, Sommerville 2011, Wiegers 2003].

Since most requirements efforts use a natural language style to specify requirements and natural languages lack the precision and rigor of formal languages, great care must be taken to ensure that natural language requirements are clear, concise, correct, consistent, verifiable, etc. In the following list, we offer a set of guidelines on writing good natural language requirements.

1. Write requirements in complete sentences in an "active voice".
   - *Example*: Write "The TMS shall provide the capability for a Traffic Manager to set the duration of individual traffic signal states, …" rather than "The duration of individual traffic signal states may be set by the TMS Traffic Manager, such that …".
2. Be precise, avoid ambiguity.
   - *Example*: The statement "The TMS shall have high reliability" is imprecise. The statement "The TMS software subsystem shall not fail more often than once in every 20,000 hours" is clear about what is meant by reliability.

Table 5.2 Software Requirements Styles

Natural Language	*Structured Hierarchical Style* – a structure such as in Figure 5.5 is used with requirements numbered and written in a natural language
	Use Case Scenario – a description of a sequence of user-system interactions as described in Figure 5.4
	User Stories – one or more sentences in the everyday language of the user of a system, which captures a requirement for a system. Used in agile processes such as Scrum and XP
Graphical Notation	*Object-Oriented Analysis Diagrams* – UML diagrams such as Use Case Diagrams, Class Diagrams, Sequence/Collaboration Diagrams
	Structured/Functional Analysis Diagrams – Data Flow Diagrams, Entity Relationship Diagrams
	Other Diagrams – Context Diagram, State Diagram, Activity Diagram, User Interface Diagram
Mathematical Notation	*Formal Specification Languages* – languages, such as Z and VDM, which use set theory and mathematical logic to specify requirements
	Object Constraint Language – UML notation which uses mathematical logic to specify properties of object classes
	Mathematically Based Diagrams – diagrams with a mathematical foundation, such as state diagrams and Petri Nets

3. Requirements need to be verifiable.
   - *Example*: The statement "The new TMS system shall result in more efficient traffic flow in the intersection" is not testable. The statement "The new TMS system shall reduce the overall wait time at the intersection by 15%" is more precise and easier to test.
4. Keep requirements concise, avoid wordiness.
   - *Example*: Write "A TMS Manager shall be able to initiate video recording of traffic at a designated traffic signal" instead of "If it is desired to perform a video recording of traffic at a traffic signal, a TMS Traffic Manager designates a traffic light and then starts the video recorder for that traffic signal and stops it when she is finished".

5. Write requirements in a consistent fashion.
   - *Example*: Use the same names for the same objects – in the TMS SRS, do not refer separately to "Traffic Light", "Traffic Signal", "Stop Light", or "Traffic Control Device", if they are synonyms.
6. Each requirement statement represents a single requirement, avoid compound statements.
   - *Example*: The statement "The TMS Traffic Manager shall be able to view traffic at a designated traffic signal or request a TMS report" represents two separate requirements.
7. Avoid duplication.
   - Repeating the same requirement at different places in an SRS can cause confusion and create an inconsistency, if the requirement is reworded or changed (in one place and not where it is repeated).

EXERCISE 5.4: WRITING SOFTWARE REQUIREMENTS

After the DH Team had analyzed the motion sensor feature and revised the Use Case Model to include motion sensor functionality, Michel and the team reviewed the *DigitalHome Software Requirements Specification*, Version 1.2 (see Appendix B), to determine what changes had to be made.

The exercise, described at the end of this chapter, deals with how the Team will revise the SRS to include the motion sensor feature.

Validating Requirements

While working at Volcanic Power, Georgia Magee had worked in the Quality Assurance Department and been part of several teams responsible for requirements validation and verification (V&V). Disha asked Georgia to lead a team discussion about the Requirements Validation Process. Georgia agreed, but asked for a few days to prepare a presentation and a discussion format.

A few days later the DH Team met, and Georgia started with some basics about verification and validation – she stated: "I know you may all be familiar with this material, but I thought it would be good to review and refresh". (Georgia's presentation and her discussion material are included in the below mini-tutorial.)

After Georgia's presentation and the team's questions, Georgia discussed the Digital Home activities, outlined in Table 5.1, and which are explicitly related to requirements V&V: Inspect and revise the SRS, develop a System Test Plan, and Develop a Requirements Traceability Matrix (RTM). The team began work on these tasks (see Chapter 3).

MINITUTORIAL 5.6: SOFTWARE
REQUIREMENTS VALIDATION

Massood Zewail remarked that he was not sure about the difference between the terms "validation" and "verification" (V&V). Georgia was ready for this with a slide. She said although there are differences in the way people use the two terms and there is often confusion, the *IEEE Standard Systems and Software Engineering Vocabulary* [IEEE 24765] defines the terms as listed on the slide:

- *Validation* – confirmation, through the provision of objective evidence, that the requirements for a specific intended use or application have been fulfilled. In colloquial terms, validation is concerned with "building the right product".
- *Verification* – the process of evaluating a system or component to determine whether the products of a given development phase satisfy the conditions imposed at the start of that phase (e.g., the design documents cover all the requirements in the SRS). In colloquial terms, verification is concerned with "building the product right".

Georgia pointed out that the *IEEE Standard for System and Software Verification and Validation* [IEEE 1012] states that the purpose of software requirements V&V is "to assure the correctness, completeness, accuracy, testability, and consistency of the system software requirements". [IEEE 1012] lists a variety of activities used for Software Requirements V&V: Requirements Evaluation, Interface Analysis, Traceability Analysis, Criticality Analysis, Software Qualification Test Plan, Software Acceptance Test Plan, Hazard Analysis, Security Analysis, and Risk Analysis.

Four common V&V requirements activities are as follows:

- *Requirements Reviews* – the developers and other stakeholder review the requirements to ensure their quality – find problems and correct them.
- *Prototyping* – a prototype can provide an early view of the "look and feel" of system and allows customer and users to validate the requirements elicited by the developers.
- *System Test Design* – preparation of test cases during the requirements phase helps in assessing whether requirements are written properly – clear, correct, and concise.
- *Requirements Traceability Matrix* – the RTM documents the traceability of each requirement: traced backward to sources and forward to software artifacts developed from the requirements. The initial construction of the RTM verifies the source of each requirement; and as it is updated throughout the development process, it supports verification of the other development artifacts (e.g., design components, code, test cases).

Chapter 3 contains additional material about V&V activities performed throughout the development life cycle.

EXERCISE 5.5: SECURITY REQUIREMENTS

After the DH Team added motion sensor requirement to SRS 1.2, Disha Chandra asked the team's thoughts about the Security Requirements in Section 5.4 in SRS 1.2.

The exercise, described at the end of this chapter, deals the team's review of Section 5.4 in SRS 1.2.

Case Study Exercises

EXERCISE 5.1: SPECIFICATION OF A QUALITY ATTRIBUTE SCENARIO

SCENARIO

After Yao Want, the DH System Architect, had completed his presentation on Quality Attributes, he led the team in specifying a quality attribute scenario for the following security requirement for the *DigitalHome* system: "5.4.2 Log in to an account shall require entry of an account name and a password".

LEARNING OBJECTIVES

Upon completion of this module, students will have increased ability to:

- Describe and discuss the nature and value of Quality Attribute Scenarios.
- Develop scenarios for security quality attributes.

EXERCISE DESCRIPTION

1. This exercise follows a lecture on design concepts and principles that were presented in MiniTutorial 5.2.
2. As preparation for this exercise students are required to read the following:
   - DH Beginning Scenario
   - *DigitalHome* Need Statement
   - *DigitalHome Software Requirements Specification*
3. Students are grouped into teams of 4 or 5
4. Each team specifies scenario for the DH security requirements 5.4.2 using the below format:
   - *Stimulus* – the actual event that the system must respond to
   - Source of Stimulus – the entity responsible for generating the stimulus
   - Environment – the conditions under which the stimulus occurs (different environments might result in different system responses)
   - Artifact – the part of the system being stimulated. This could be the whole system, or a single artifact (e.g., source code, a test plan, …)
   - Response – the desired activity that must be performed, because of the stimulus
   - Respo*nse Measure* – a way of making the response quantifiable
5. The team chooses one member of the team to make an oral report to the class on the team's recommendations.

EXERCISE 5.2: ELICITING DH REQUIREMENTS

The DH Team decided to carry out a formal elicitation effort. This exercise deals with how the DH Team will plan for its elicitation activities.

LEARNING OBJECTIVES

Upon completion of this module, students will have increased ability to:

- Describe various requirements elicitation methods.
- Plan for elicitation of system and software requirements.
- Use elicitation techniques to collect requirements.

EXERCISE DESCRIPTION

1. As preparation for this exercise, read Chapter 1 and this chapter.
2. You will be assigned to a small development team (like the DH Team).
3. Your team is to take on the role of the DH Team and organize and plan for DH Requirements Elicitation, carrying out the following tasks:
   a. Review the *DH Development Process* script and the HLRD. Discuss the need for information about DH Requirements, making a list of key parts of the HLRD that need additional detail.
   b. Address each of the following elicitation questions:
      - Which elicitation technique(s) should be used? What is the rationale for the selection?

- What pre-work must be done before applying the technique?
  - Who will act as users/customers?
  - Who are other stakeholders for the DH system?
  - What elicitation instruments will need to be prepared?
  - Who from the DH Team will carry out the pre-work?
  - How long will the pre-work take?
- What is the task plan for the actual elicitation work?
  - What are the elicitation tasks?
  - Who, from the DH Team, will perform the elicitation tasks?
  - How many hours, over how many days are estimated to carry out the elicitation?
  c. Answer the questions in writing, using the below Plan Template, and choose one member of the team to report to the class on the group's conclusions.

Software Requirements Elicitation Plan

Pre-Work Tasks	Decisions/Estimates	
Areas of the HLRD needing additional information		
Elicitation Technique(s) to be used with a rationale statement for the selection		
List the potential elicitation users/customers		
Elicitation instruments to be prepared		
DH Team members(s) responsible for Pre-Work		
Estimated number of hours for Pre-Work		
Elicitation Task	**Responsible DH Team Member**	**Estimated Hrs**

EXERCISE 5.3: MODELING A DH USE CASE

It has been decided to add a motion sensor feature to the *DigitalHome* prototype. The DH Team will conduct an analysis of the motion sensor feature and revise the Use Case Model, adding the feature.

This exercise deals with the Team's analysis of the motion sensor feature, including use case modeling of the feature.

LEARNING OBJECTIVES

Upon completion of this module, students will have increased ability to:

- Perform domain research
- Develop a use case diagram
- Develop a use case scenario

EXERCISE DESCRIPTION

1. As preparation for the case module, read Chapter 1 and this chapter, and review the DH Use Case Model, Version 1.0 (Appendix C).
2. You will be assigned to a small development team (like the DH Team, 2–3 people). Each team analyzes and models the addition of the motion sensor feature to the DH requirements:
3. Each team analyzes the motion sensor feature:
   - Conduct domain research about motion sensors: how they work; what are their capabilities and limitations; what model(s) would be appropriate for the DH system; etc.
   - Determine what characteristics of the feature should be present in the DH system: what type of sensor, devices (e.g., lights) to be controlled; what will be the time delays for "on" and "off" conditions; how will the user interact with the motion sensors; etc. This may require interaction with potential users.
4. Determine what changes and additions must be made to the DH Use Case Model, Version 1.0, with the addition of a motion sensor feature.
   - Determine any changes that need to be made to the DH Use Case Diagram.
   - Determine changes that need to be made to any use case scenarios.
   - Develop any new use case scenarios needed for the motion sensor feature.
5. Document your changes in DH Use Case Version 1.0. Mark up changes on the DH Use Case Version 1.0 and complete the below Use Case Description Template for any new use cases.
6. Choose one member of the team to make an oral report to the class on the team's recommendations.

Use Case Template

Use Case ID: *<Use Case ID, e.g., UC-1>*
Use Case Name: *<e.g., Customer Deposits Money>*
Goal: *<subject-verb-object format, the goal to be achieved by the use case>*
Primary Actors: *<external entity who has goals/needs satisfied by the UC, e.g., Customer>*
Secondary Actor: *<external entity indirectly involved in achieving the use case goal>*
Pre: *<a pre-condition that must be true before carrying out the use case scenario>*
Post: *<a post-condition that must be true after the completion of use case scenario>*
Main Success Scenario: *<describes a set of scenario steps which will achieve the uses case Goal>*

Step	Actor Action	Step	System Reaction

UC GUIs: *<GUIs referenced in the scenario>*
Exceptions (Alternative Scenarios of Failure): *<describes what happens when an exceptional situation occurs, which would interrupt the Main Success Scenario and require and alternate scenario>*
Use Cases Utilized: None
Notes and Issues: *<description of supporting actors, concurrency of actions, or any additional information such as special constraints or non-functional requirements related to the UC>*

EXERCISE 5.4: WRITING SOFTWARE REQUIREMENTS

After the DH Team had analyzed the motion sensor feature and revised the Use Case Model to include motion sensor functionality, the team reviewed the *DigitalHome Software Requirements Specification*, Version 1.2 (Appendix B), to determine what changes had to be made.

This exercise deals with how the team will revise the SRS to include the motion sensor feature.

Upon completion of this module, students will have increased ability to:

- Describe the organization and content of an SRS.
- Write requirements that are precise and unambiguous.
- Specify requirements in an SRS document.

EXERCISE DESCRIPTION

1. Exercise 5.3 on Use Case modeling of the motion sensor feature should be completed first.
2. As preparation for the case module, ask each student to review the DH SRS, Version 1.2.
3. You will be assigned to a small development team (like the DH Team). Each team acts as the DH Team in specifying the addition of the motion sensor feature to the DH requirements:
4. The team specifies the motion sensor feature by including the motion sensor feature in DH SRS 1.2.
   - *Determine which parts of the SRS 1.2 require change or addition.*
   - *A*ssign individual team members tasks of revising the various parts of SRS 1.2.
   - Review the changes and additions by team members to ensure that the motion sensor feature is properly specified and no other features or defects have been introduced.
   - Compile a report of changes and rewritten or new requirements.
5. Choose one member of the team to make an oral report to the class on the team's SRS work.

EXERCISE 5.5: SECURITY REQUIREMENTS

After the DH Team added motion sensor requirement to SRS 1.2, Disha Chandra asked the team what the thought about the Security Requirements in Section 5.4 in SRS 1.2. She had concerns about whether the security requirements were covered sufficiently.

The exercise deals the team's review of Section 5.4 in SRS 1.2.

LEARNING OBJECTIVES

Upon completion of this module, students will have increased ability to:

- Describe the importance of security requirements.
- Identify requirements that are needed to ensure the security of a system.
- Specify security requirements in an SRS document.

Exercise Description

1. As preparation for the case module, ask each student to review the DH SRS, Version 1.2.
2. You will be assigned to a small development team (like the DH Team). Each team acts as the DH Team in reviewing the DH security requirements in Section 5.4 of SRS 1.2.
3. The team carries out the following tasks:
   - Determines which parts of the SRS 1.2 require change or addition.
   - Assigns individual team members tasks of revising Section 5.4 of SRS 1.2: changing existing requirements or adding new requirements.
   - Reviews the changes and additions by team members to ensure that they properly specified and no other features or defects have been introduced.
   - Compiles a report of changes and rewritten or new requirements.
4. Choose one member of the team to make an oral report to the class on the team's SRS work.

Chapter 6

Designing *DigitalHome*

Software Design Concepts and Principles

On 11/1/202X, the DH development team, after establishing what the team felt was a solid set of requirements as specified by the DH software requirements specifications (SRS), held a team meeting to discuss the next step in the development process. In the meeting, Team Leader Disha Chandra made the point that the *Homeowner* organization had learned from previous project failures that they needed to have a greater emphasis on the design phase of the DH project. Disha asserted that a lack of effort in this phase typically results in major issues in meeting project deadlines and budgets, as well as a lack of overall quality in the system produced.

During the meeting, Disha asked Yao Wang, the System Architect, to give an overview of the design process and some of the challenges typically encountered in such a process.

Yao started with a description of the DH design phases, listed in Table 6.1 (extracted below from Table 2.2 in Chapter 2):

■ First, they would develop a software architecture for the *DigitalHome*.
■ Then, they would use the architectural model to develop a plan for building and integrating the increments of the system.
■ Finally, they would construct each increment and integrate it into the system.

Yao announced that he would be delivering several short tutorials on software design. He stated he was a little wary about this since the team had extensive design experience; but he felt it would be good to review some basics and emphasize key concepts – heads were nodding as he spoke.

Table 6.1   DH Project Design Phases

Architectural Design	• Develop an Object-Oriented (OO) high-level design model
	  • Refine the OO analysis model
	  • Develop additional design models
	  • Specify and document design in an SDS
	• Develop a Build/Integration Plan
	• Review and revise the SDS
	• Consult with customer, and revise SRS and SDS, as appropriate
	• Track, monitor, and update plans (Task/Schedule/Quality/CM/Risk) & RTM

Construction Increment 1, 2,...	• Complete Class Specifications (data types and operation logic) • Develop incremental Integration Test Plan (Class/Integration) • Set up an appropriate test environment (e. g., test harness, stubs/drivers) • Review the Class Specifications • Write source code • Review and test the source code • Integrate code with previous increments • Perform Increment (Class/Integration) Test • Consult with customer, and revise SRS, SDS, and code, as appropriate • Track, monitor, and update plans (Task/Schedule/Quality/CM/Risk) & RTM

MINITUTORIAL 6.1:   SOFTWARE DESIGN CONCEPTS AND PRINCIPLES

The purpose of software design is to produce a workable (implementable) solution to a given problem [Budgen 1989]. The SWEBOK [Bourque 2014] states

> A software design (the result) describes the software architecture— that is, how software is decomposed and organized into components—and the interfaces between those components. It should also describe the components at a level of detail that enables their construction.

According to [Booch 1999]: a design component is "an abstraction of reality that helps us understand complex systems and provides a blueprint for a system implementation, and should omit minor elements that are not relevant to the level of abstraction of the module".

Note: The terms "module" and "component" are sometimes used interchangeably when discussing the elements of an architecture. In this discussion, we will use the term *module* to designate a design-time entity (e.g., a package or an object class), and we will use the term *component* to designate a run-time entity (e.g., an executable binary).

Software design represents a key challenge within the software development process. The software design must be developed in a way to allow for seamless implementation of the modules defined during the design. The design should also allow for easy maintainability of the system in case of a future change, which is based on changes in the requirements or because of a discovery of a defect within the system. These factors can be achieved by following a set of well-known design principles:

MODULARITY

Modularity in design refers to dividing the system into smaller modules or parts, such that each module contains a subset of the functionality of the overall system. Well-designed systems are made up of modules that can be independently created, tested, and maintained, and possibly be reused in future systems. Modules interact with each other through their interfaces. Modular design allows developers to break down a complex problem into a set of smaller, less complex, and easier to solve problems.

Figure 6.1 depicts how, at a high level, a design for a Traffic Management System (TMS) could be divided into modules.

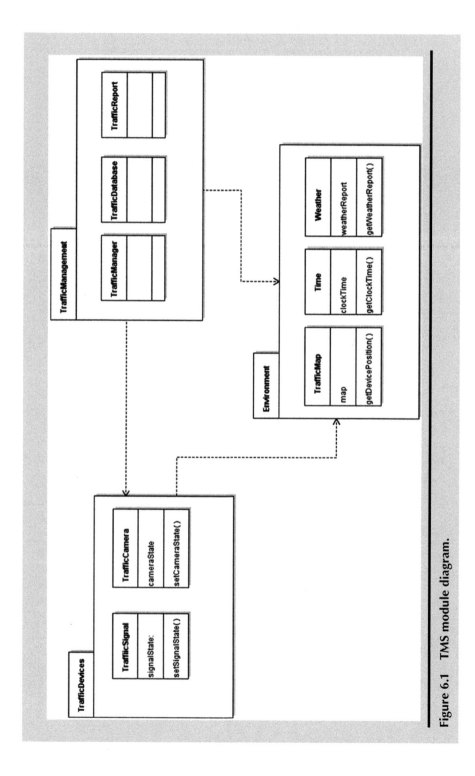

Figure 6.1 TMS module diagram.

ABSTRACTION

Abstraction is an approach that focuses on the overall purpose of a design element, without being distracted by low-level details. Modularity supports using abstraction to allow the designer to think of the system in terms of black boxes (modules or components) that deliver specific functionality and have interfaces for communication, without focusing on the implementation details of these boxes. Figure 6.1 might be the first level of design of the TMS. It hides the details of the implementation of the modules (packages and classes) and their dependency links (dashed lines). Such details would be determined in lower level design activities.

In object-oriented design for example, one might define a class of objects that has certain functionality and leaves the implementation of such functionality for a later time. Abstraction is used at various levels: module/component abstraction, procedural abstraction, and data abstraction. Two key concepts underlying the use of abstraction are "encapsulation" and "information hiding".

ENCAPSULATION

We say that an object *encapsulates* its state and behavior – that is, an object is treated as a single entity with the details of its implementation hidden from other objects that interact with it. With this approach, a software system class is viewed as made up of a set of objects that model elements of the software's problem domain. The execution of the software is characterized as follows:

- The software causes the objects to be created with some initial state.
- The objects communicate with each other by sending messages to each other.
- A message requests information about an object or it may change the state of an object (or create or destroy an object).
- Through this transmission of information and change of state, the software can satisfy the software requirements.

Figure 6.2 shows a class diagram for the TrafficManagement package. The classes are depicted at various levels of abstraction: the Set class only specifies the name of the class, while the TrafficManager class specification has more detail about class attributes and methods. The TrafficManager class encapsulates the state of the class in the top half (ID, PIN,) and the behavior in the bottom half (login(), loglout(), sekect(), etc.). *Note*: the+sign next to a method means it has "public" visibility (it is accessible by objects outside the class); and the − sign next to an attribute means it has "private" visibility (not accessible by objects outside the class).

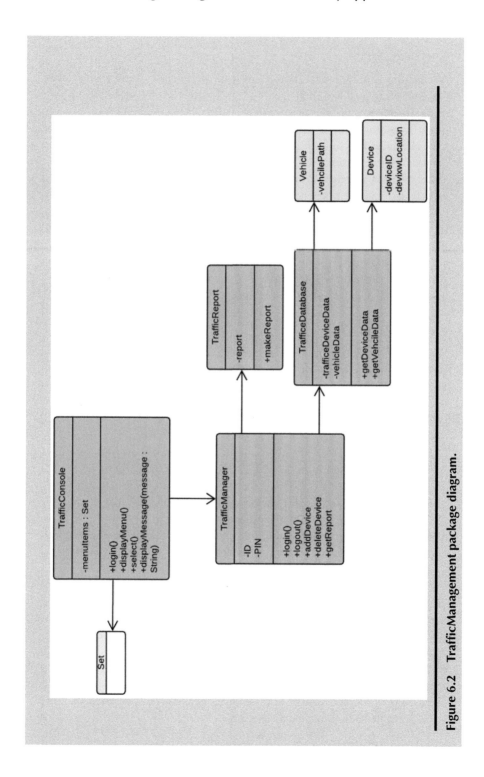

Figure 6.2   TrafficManagement package diagram.

INFORMATION HIDING

One of the most important ideas in object-oriented software design is the idea of information hiding [Parnas 1972]. *Information hiding* is a design principle that limits access of the user of an object by constructing a class interface that acts as a communication medium between the user of the class and the class. The interface hides the details of the class implementation, with the effect that the complexity of the design is reduced, the maintenance capability is enhanced, and the security of data is protected. Information hiding allows encapsulation of data and behavior so that an object is viewed as a single entity.

The object interface provides controlled access to an object's data elements and a means for communicating with and performing operations on an object. Implementation languages such as C++, Java, and Python provide the capability to construct an interface for an object that provides basic information about the object, represents the methods that can be applied to the object, and hides the details of the object implementation.

In Figure 6.2, the data element "menuItems" is declared as a Set type; but, the details of Set are "hidden" in a separate class. The details of the method "displayMenu()" are not shown; hence, if changes are made in Set it will not require changes in the software using these features.

HIGH COHESION

Cohesion is a measure of the degree to which the elements of a module are functionally related [Stevens 1974]. A highly cohesive module implements functionality that is related to one feature of the solution and requires little interaction with other modules. For example, in OO design, a highly cohesive class is one where all methods present in that class are included because they provide a subfunction of the major function of the class. Notice in Figure 6.2 that all the TrafficManager methods are related to operations that a traffic manager would perform. Similarly, in a system composed of multiple packages (as in Figure 6.1), each class within a package contributes to the overall purpose of the package. Good design incorporates strong cohesion of modules.

The level of cohesion of a module ranges from functional cohesion (highest level of cohesion) to coincidental (lowest). In functional cohesion, all submodules or methods within a module are related and contribute to the overall goal of the module. In coincidental cohesion on the other hand, as the name indicates, the submodules are not related and seem to be grouped together for no obvious purpose. Such modules are hardly ever reused outside of the current system.

LOW COUPLING

Coupling is a measure of the strength of the connections between the different system modules [Stevens 1974]. Systems made of modules that strongly rely on each other are said to be tightly coupled. On the other hand, loosely coupled systems are made up of modules which are relatively independent. In Figure 6.1, there would be greater dependence between the classes in the TrafficManagement Package than between the classes in the Environment package. We would say the coupling is "tighter" in the TrafficManagement packages. Looser coupling between modules mean there is more independence between them; hence, a change in one module would have less impact on the other.

Like with cohesion, coupling between modules has ranges. *Data coupled* modules exchange only the data items that are necessary, and they do not depend on the implementation of the elements, which is the lowest and most desired level of coupling. For example, in Figure 6.2, the TrafficManager does not depend on how TrafficDatabase and its attributes are implemented (of course, the methods getDeviceData and getVehcileData do depend how the attributes are implemented). In *content coupling*, modules reference the content (data structures and implementations) of other modules.

SECURITY

Security is a measure of a system's ability to protect itself from attacks meant to harm the system. Three characteristics of such protection are as follows [BASS 2012]:

- *Confidentiality* – the property that data or a service is protected from unauthorized access. For example, an illegal user might try to change the ID of a TrafficManager object in the TMS.
- *Integrity* – the property that data or services are not subject to unauthorized manipulation. For example, someone may try to deliberately alter data in a TafficReport in the TMS.
- *Availability* – the property that the system will be available for legitimate use. For example, the access of a legal user to the TMS cannot be denied.

The previously discussed Design Concepts and Principles support building a secure system. Modularity, abstraction, encapsulation, high cohesion, and low coupling help provide a design where the essential features of the system design are well organized and clearly defined, without the detail that might inhibit envisioning security problems. Information hiding limits and provides access of a user to only those elements needed by the user.

EXERCISE 6.1: UNDERSTANDING DESIGN CONCEPTS AND PRINCIPLES

After Yao Wang had completed his presentation on Design Concepts and Principles, he opened discussion with a set of questions about the topic. The exercise, described at the end of this chapter, deals with how the Team responded to the questions.

Software Architecture

After Yao Wang delivered his introductory tutorial on design concepts and principles, he explained that while defining a software architecture that ensures addressing the functional requirements (i.e., that the system does what it is supposed to do), it is just as important that the system architecture ensures the inclusion of the non-functional requirements (quality attributes) of the system such as performance, reliability, security, availability, testability, and modifiability. Yao also explained that architecture differs from lower-level design in the sense that architecture tackles the overall structure and behavior of a system as bounded by a set of software requirements, while lower-level design decisions are those made by developers to design system modules that maintain the overall architectural decisions made earlier in the architecture phase.

Next, Yao prepared and delivered a tutorial on development of a software architecture.

MINITUTORIAL 6.2: ROLE OF SOFTWARE ARCHITECTURE

The activities associated with arriving at a sound architecture and design for a software system are critical for the long-term well-being of the developed system. Similar to the requirements phase, a lack of emphasis on the architecture of the system can lead to future issues related to budgeting, scheduling, maintenance, and the quality of the system under development.

Software architecture is still considered a relatively new field and as such there is still not a universally agreed upon definition of what software architecture exactly is or is not. Indeed, the Software Engineering Institute website lists over one hundred definitions of software architecture (http://www.sei.cmu.edu/architecture/start/glossary/community.cfm). These definitions come from experienced software architects, analysts, managers, researchers, and software engineers in general. Although there exists a wide variability in the definitions, there is a common theme to all of them: developing a *software architecture* is concerned with decomposing a system into multiple modules that function and collaborate to ensure the overall functional requirements and quality attributes of a software system. This is not too dissimilar from the architecture for an office building: such an architecture would include a description of the building modules (the structure of the foundation and floors, the electrical system, the plumbing system, the air-handling system, etc.) and how the modules function and collaborate. Like a building architecture, software architectures are typically described, at their highest level, with a diagram of a set of modules and their connectors (interactions), such as in Figure 6.1.

Architecture employs the notion of abstraction in the definition of these structures/modules to avoid the focus on implementation details such as algorithms, and data representation, and instead treats these modules as black boxes with responsibilities. Each structure must have a well-defined interface and protocols for collaborations. It is important to note that when talking about structures, we refer to entities that ultimately represent a piece of the software under development. This includes subsystems, layers, modules, packages, and classes. Although there is a focus on the notion of breaking the system into modules or structures, software architecture also emphasizes behavior of these modules. Behavior within a software architecture focuses on describing the behavior of each of the modules, as well as, the way these modules interact together.

The question arises; how does a software architecture come about? In breaking a system into multiple collaborating modules, we typically map the set of required system functionality into one or more modules that are responsible for carrying out these functionalities. However, while ensuring completeness and correctness of system functionality is a major concern in developing an architecture, quality attributes are also a major factor in the development. Quality attributes deal with those constraints under which the system must be developed or execute. These constraints are organizational, environmental, developmental, or ones related to the execution of the system. Examples of organizational constraints include budgets and deadline for the development of the system. Constraints associated with the environment with which the software system interacts include the available hardware that the software will utilize and the type of users or other systems that interact with the developed system. The development team's skill and competency is an example of a developmental constraint.

Identifying and consolidating all these concerns is a principal goal of software architecture. Software architecture must address the concerns of many stakeholders of the system. A typical set of stakeholders might include some of the following: customers, users, managers, marketing personnel, developers, quality assurance personnel, maintenance personnel, and government regulators. Most likely, different stakeholders will have conflicting requirements. For example, the maintenance stakeholders desire a system that is easy to modify and correct, while an end user would like a system that is secure, reliable, and easy to use. This is the major challenge of developing software architecture; mutually satisfying a vast number, and in many instances, a conflicting set of functional requirements and quality attributes (requirements related to the performance or use of the software).

As the architecture is being developed and architectural decisions are being made, it is important to document the architecture in a way that clearly describes how it addresses the needs of all the respective stakeholders. The document must be written and organized in a way that can be understood by each stakeholder.

In general, an architecture document can be used to:

- guide the construction of the system through explicitly declaring architectural decisions that must be adhered to during detailed design and implementation;
- introduce new people (developers, analysts, customer,...) to the system;
- communicate with stakeholders;
- and perform analysis to ensure the attainment of requirements.

Architecture Views and Styles

After specifying the set of system requirements in the SRS document and arriving at a set of quality attribute scenarios, the next step for the DH development team is to establish the path for moving from the requirements phase to establishing a system architecture. System Architect Yao Wang called for a meeting where Wang expressed the importance of developing and documenting a system architecture that can be useful for the multiple stakeholders of the system. In his listing of those stakeholders he included:

- Customers
- Users
- Marketing personnel
- Managers
- Development team members
- Maintenance personnel
- Independent V&V personnel

Yao asked for suggestions on how to proceed with representing an architecture that could be used by all these entities, who have different sets of expertise, technical knowledge and language, and varying concerns. Georgia Magee suggested that the team come up with a priority of the stakeholders and address the needs for the top three. This was not acceptable by the team lead. Disha Chandra made the point, that since this is a prototype system, the concerns of all stakeholders must be adequately addressed and represented. Michel Jackson suggested that the team come up with a very general architecture that could be understood and used by all stakeholders. This suggestion was also rejected. Yao Wang indicated that an architecture must be detailed enough to ensure correct modular and detailed design and implementation by downstream developers. Finally, Yao suggested that the team use what is known as "architecture views" and "architectural styles" to represent the system in different ways, each of which is suitable for a set of stakeholders. Since not all the team was familiar with architectural views and styles, he offered to prepare and present a mini-tutorial on the subject.

MINITUTORIAL 6.3: ARCHITECTURE VIEWS

As mentioned earlier, software architecture addresses the needs of multiple stakeholders with varying technical expertise, technical languages, and concerns. To establish an architecture that considers a wide range of concerns, the architect must look at the system through different lenses; the architect should look at the system from the different perspectives of the customer, users, developers, and managers, among others.

Architectural views are representations of the overall architecture that are meaningful to one or more stakeholders in the system. The architect selects a set of views that allows the architecture to be communicated to and understood by all the stakeholder; and it enables them to verify that the system will address their concerns. Views allow architects to use different perspectives to break the system into modules and make architectural decisions. It is important to note that the information included in different views is complementary, not contradictory. Views can be tackled concurrently if control is provided to ensure the consistency among the information or decisions made in each of the views.

The literature contains a wide range of architectural views. In this tutorial, we look at three of the most common views: Kruchten's "4+1 views" [Kruchten 2004], the Software Engineering Institutes (SEI) "Views and Beyond" approach [Clements 2010], and the "Siemens" approach to architectural views [Hofmeister 2000].

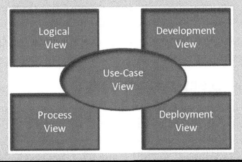

Figure 6.3   4+1 Architecture views.

KRUCHTEN'S 4+1 VIEWS

The Krutchen's approach to software architectural views proposes a model made up of four main views plus an additional view that ties the other four together.

These views are illustrated in Figure 6.3 and described as follows:

- *Logical View* – provides an object model of the design where Object-Oriented Design is the design paradigm used. The logical view focuses on mapping the functional requirements of the system to objects or object classes. The notation mostly used with this view is a UML (Unified Modeling Language) class diagram [Fowler 2004].
- *Process View* – Focuses in tackling the issues of concurrency and synchronization within the architecture. The process view concentrates on quality attributes such as performance and availability. In this view, the software elements of interest are tasks which represent the system's runtime threads of control mapped to computing processing nodes.
- *Deployment View (or Physical View)* – Introduces the hardware aspects of the architecture and shows how software elements map to hardware elements. Mapping of the tasks defined in the process view to a physical hardware is the focus of the deployment view. This view also addresses quality attributes such as performance, reliability, availability, and scalability.

■ *Development View* – Focuses on the development team and organization and shows how software elements are assigned and managed by the team and management. The development view has a project management focus that breaks the system into subsystems that can be assigned to a set of developers. This view specifies the set of skills required for each subsystem and the amount of time required for developing each subsystem.

■ *Use Cases View (Scenarios)* – This is the view that combines the other views together. Decisions made in the other four views are motivated by the scenarios of the uses of the system. The Use Cases View provides the set of use case scenarios that support the decisions made in the other four views. These scenarios ensure consistencies among the other views and they provide a way of quantifying and validating the resulting architecture against a common set of scenarios.

SEI'S VIEWS AND BEYOND APPROACH

The SEI's Views and Beyond Approach (V&B) defines a set of three architecture views, by which a system's architecture can be defined [Clements 2010]. While the views and design decisions in 4+1 approach were derived from a set of system use case scenarios, the SEI approach relies on the notion of quality attribute scenarios to determine the appropriate views and architectural decisions. The three views defined by this approach are as follows:

■ *Module View* – Focuses on defining system modules that have responsibilities as part of the system functionality. In other words, this view focuses on mapping functional requirements of the system into a set of responsible modules.

■ *Component and Connector (C&C) View* – Focuses on defining the runtime software elements. In this view, modules defined by the module view are mapped into runtime entities such as processes or threads. For example, a class in the module view might be mapped to a set of objects available at runtime. Since this view focuses on the runtime behavior of the system, it is used to analyze quality attributes of the system such as performance, availability, and security.

■ *Allocation View* – Like the Development view in the 4+1 approach, this view relates software elements defined in the previous two views to the environment under which the system will be developed or will interact with. In this view, software elements are mapped to hardware elements such as processors or clients and servers, or to an organizational or work unit such as an individual developer, a team of developers, or a subcontractor who will be responsible for the development of that module.

These views are illustrated in Figure 6.4.

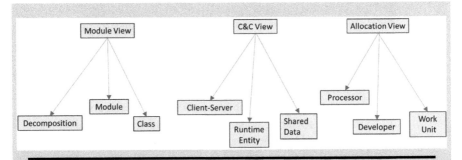

Figure 6.4   SEI's V&B approach.

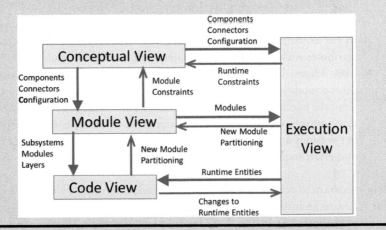

Figure 6.5   Siemens approach.

SIEMEN'S APPROACH

The Siemens Four-Views approach [Hofmeister 2000], developed at Siemens Corporate Research, uses four views to document an architecture. While the 4+1 approach relies on use case scenarios and the SEI approach uses quality attributes scenarios to derive architectural decisions, the Siemen's approach uses the notion of global analysis to ensure the identification of stakeholders needs and concerns, and to derive the architectural design within each of four views. The four views of the Siemen's approach are pictured in Figure 6.5 and described as follows:

■ *Conceptual View* – Maps the system's functionality to a set of decomposable, interconnected modules and connectors. The conceptual view's primary concern is ensuring that system functionality is mapped to implementable components. Decisions such as development of in-house modules or relaying on commercial-of-the-shelf (COTS) are addressed in this view.

■ *Module View* – Breaks the system into a decomposition structure and a set of layers. The decomposition structure presents the system's logical decomposition into subsystems and modules, and each of the resulting modules is assigned to a layer. The primary concerns of the module view are to minimize module dependencies, to allow for module reuse, and to address the impact of future changes in elements related to the system such as COTS software, operating system, hardware, and other external software.

■ *Execution View* – Describes the system's structure in terms of its run-time elements such as tasks, processes, and threads. The tasks associated with the execution view relate to assigning system functionality to run-time hardware elements, determining how the system runtime entities communicate with each other, and the allocation of physical resources to the executing elements. As such, this view addresses system quality attributes such as performance, safety, reliability, and availability.

■ *Code View* – Focuses on the organization of the software artifacts into source modules, from the module view, and deployment components, from the execution view. The code view is concerned with providing version control support for the software artifacts as well as controlling product versions and releases. This is an important view for managing time and effort required for product updates, build time, and integration.

It is important to note that while there exist a wide number of view approaches for software architecture, the set of views provided typically contain overlapping elements. As in the 4+1's logical view, the SEI's Module view, and the Siemen's Conceptual view all aim at addressing the functional requirements of the system while the other views tackle quality attributes concerns. Choosing a view approach depends largely on the system being developed, the set of system stakeholders and their concerns, and the environment under which the system will be developed, used, and maintained.

EXERCISE 6.2: SELECTING APPROPRIATE SET OF VIEWS FOR THE DH SYSTEM

After the DH Team had developed a set of system requirements for the DH, it was now time for the team to focus on developing the system architecture. System Architect Yao Wang suggested that the team review the DH artifacts that identify the system stakeholders and their needs to clearly identify and group the different stakeholders. Then next, determine a set of architectural views that could be used to document the overall system architecture. The exercise, described at the end of this chapter, deals with how the Team will arrive at the set of appropriate views for the DH system.

MINITUTORIAL 6.4:   ARCHITECTURE STYLES

Although Architecture Views provide a way for viewing the system from different perspectives, they are not intended as concrete solutions to a design problem. Views such as the Module view in the SEI approach or Krutchen's Process view give overall guidelines of the type of information that must be present in the view, but they do not provide a way of breaking the system into components and how these components interact and relate. This indeed is the purpose of *Architecture Styles*.

An *architecture style* represents a widely observed and proven solution to a recurring design problem. A style provides "a pattern of structural organization" [Shaw 1996]. The same solution (style) can be applied to completely different systems. Styles provide the ability to apply specialized design knowledge to the design of a similar class of systems. A style provides guidelines for how different components are arranged within a design and specifies the restrictions on the interactions among components. Each style has a set of quality attributes that it intends to address.

The literature provides an abundance of architecture styles, which architects can select from to solve design problems. In the rest of this section, we discuss some of well-known styles. We describe each of the styles by defining the different types of components associated with the style, the type of relations that may exist between these elements, and the restrictions imposed on the elements and relations in the style.

PIPE AND FILTER STYLE

This style is characterized by continuous transformation of data from one component to another where the output of one component is the input of the next. The two types of elements in this style are Pipes and Filters. A *filter* is the type of element that transforms data received as input and pushes the new transformed data as output. A *pipe* serves as a connector between two filters; pipes are responsible for mitigating the transformation of data between filters. Each pipe and filter have input and output ports to allow for connections between the different elements. Each filter's output port will be connected to a pipe's input port, and a pipe's output port will be connected to at least one filter's input port. The restriction on this style is that pipes must be connected to filters and filters to pipes. In other words, between each two pipes there must be a filter and vice versa.

The advantage of this style stems from the fact that filters execute concurrently and independently: filters should not share state with each other and do not know which filters are upstream or downstream. A filter begins to work once it receives the input data it needs to begin working. Filters can pass just enough data as needed by the next filter and continue working. This parallelism obviously enhances system performance. Examples of the pipe and filters style are compilers, signal-processing systems, distributed systems, and UNIX pipes [Clements 2010]. Figure 6.6 shows a Pipe and Filter style used for a compiler:

Figure 6.6   Compiler pipe and filter.

CLIENT/SERVER STYLE

The *client/server architectural style* describes distributed systems that involve interacting components in a connecting network. The two types of components are *clients*, which request information, and *servers*, which furnish the information to the client. Clients know the type of service they need and the identity of the server providing this service. Clients are responsible for initiating the request for a server. It is possible that a single component can serve as both a client to a server and a server to other clients. The style also allows for a single server for clients or multiple distributed ones to enhance system availability in case of a malfunction in one of the servers. Figure 6.7 pictures a client-server architecture with multiple clients served by a single server.

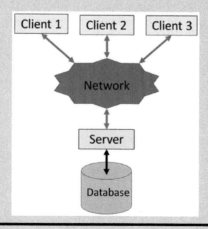

Figure 6.7   Client-server architecture.

Typical restrictions for this style are the number of clients for a particular server and whether or not a server and client can serve each other, which might cause undesired behavior due to circular dependency and possible deadlock.

A typical example of a system in the client-server style would be an information system running on local network in which GUI applications request information from a database management system acting as a server. Web-based applications are another example of this style, where the clients run on web browsers and the servers are components running on a web server.

LAYERED STYLE

The *layered style* allows for the division of the system under development into units that are stacked on top of each other. These units are called layers. A layer contains one or many software components such as packages, subsystems, or classes, and is responsible for a significant portion of the system. In a layered system, each layer provides service to the layer above and serves as client for the layer below [Shaw 1996]. However, elements within a layer can interact freely with each other.

This is one of the most common architectural styles and most complex software systems make use of this style. The main motivation behind this style is that layered systems promote modifiability and portability. Layered systems also allow for parallel development of each of the layers; and in a truly layered system, layers are highly cohesive and may be reused in future systems. As mentioned above, this is a style that is widely used and applicable to many software systems. An example of a typical layered architecture is shown in the Figure 6.8.

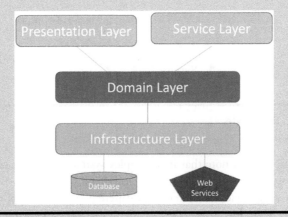

Figure 6.8   Layered style.

SERVICE-ORIENTED STYLE

In the *service-oriented architecture* (SOA) style there are two types of components; producers of services and consumers of those services. These components are usually distributed. Consumers and providers are independent: they can be implemented in different languages; built on different platform; they can be deployed independently; and may not (that is typically the case) belong to the same system. A *service* is a stand-alone unit of functionality that is made available by providers and used by consumers.

Both service *providers* and *consumers* subscribe to an intermediary entity that mitigates the publishing of a newly available service (by a provider) and the request for a service (by a consumer). A typical example of such a mitigating entity is a *service broker*. A service broker would maintain a registry of services, which are available, and would allow for communication between service consumers and service providers. Figure 6.9 depicts a general view of an SOA using a service broker.

The major attraction of this style is interoperability. Since service providers and service consumers can run on different platforms, this type of style allows for the integration of different systems which can be an effective way of incorporating legacy systems with newly developed systems.

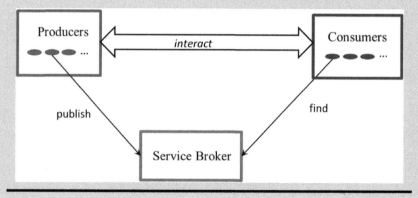

Figure 6.9   SOA style.

It is important to note that in a complex system there may be a different number of styles used. It is seldom the case that a single style will be applied throughout a multifaceted system. So, different styles are applied to different areas of the architecture. Also, a style of one type might be used within another; for example, a server in a client-server style might be implemented using a pipe-filter style.

EXERCISE 6.3 SELECTING AN ARCHITECTURAL STYLE

After identifying the architectural views to be used as part of the system architecture (Exercise 6.2), the DH Team was ready to start working on finding concrete design solutions using available architecture styles. The DH Team Leader, Disha Chandra, tasked Yao Wang (with the rest of the team) to identify the appropriate architectural styles to be used. The exercise, described at the end of this chapter, deals with how the Team will arrive at the architectural styles to be used in the DH architecture.

Object-Oriented Design

After the DH Team selected and described the architectural styles for the DH system, Yao Wang began a discussion of which design paradigm should be used to implement the styles selected. The team was in universal agreement that they should use an object-oriented design approach. Just to make sure the team had a common and accurate understanding of objected-oriented design features, Yao led a discussion of the features.

MINITUTORIAL 6.5: OBJECT-ORIENTED DESIGN

Object-oriented design is an approach in which systems are broken down into modules, which are composed of packages and object classes. An *object class* is template for a set of *objects* that contain both data and functions (methods), as in the class TrafficManager in Figure 6.2. Each object created (instantiated) has three components; (1) a unique identifier (name) which distinguishes it from other objects during system runtime, (2) a state, which is represented by the values of the object variables (attributes), and (3) a behavior, which is represented by the methods defined within an object class. For example, a *TrafficManager object* has a unique name, such as *Manager232*, to distinguish it from other TrafficManager objects available at runtime. *Manager232* might have a state with an *ID* of TBH; and it could perform functions (methods) such as adding or deleting devices into *Manager232*.

A *package* is a module that embodies a set of object classes and their dependencies. For example, the *TrafficManagement* package in Figure 6.1 contains three object classes: *TrafficMap*, *TrafficDatabase*, and *TrafficReport*. Although dependencies are not shown between the classes, we would expect that *TrafficReport* depends on *TrafficDatabase* (and others). Similarly, classes from other packages would have relationships to *TrafficMap* (e.g., *TrafficDevice* and *TrafficManager*).

UML is a general-purpose modeling language used extensively to model structure and behavior of object-oriented systems. UML was introduced by Grady Booch, Ivar Jacobson, and James Rumbaugh from Rational Software

[Booch 1999]. The language was later adopted by the Object Management Group (OMG) in 1997 and accepted as a standard for modeling software-intensive systems by the International Organization for Standardization (ISO) in 2000.

UML is a powerful language for use in software architecture and design, as it allows for the modeling of system's structure and behavior. Structure (static) modeling in UML is accomplished by using class diagrams and composite structure diagrams such as packages. The behavior (dynamic) modeling of the system or individual components is accomplished by using sequence diagrams and state machine diagrams. The remainder of this tutorial gives a brief introduction to each of these diagrams.

CLASS DIAGRAMS

A UML Class diagram consists of a set of classes and the relationships among them. For each class, the diagram shows the attributes and methods of each class along with the visibility of each of these attributes and methods. A visibility of an attribute or a method can be *public*, *private*, or *protected*, among others. A *Public* attribute or method is available throughout the system (for objects of the same class as well as others). Access to a *Private* method or attribute is restricted to members of the same class. *Protected* attributes and methods can be accessed by members of the same package or by its subclasses. Consider the TrafficManager class in Figure 6.2. It has two private attributes (indicated with the – symbol) and five public methods (indicated with the + symbol). Classes can have different types of *relationships*:

■ *Association* – a relationship consisting of links, which connect the class objects. Figure 6.10 shows the links connecting the classes. Associations can be uni-directional (using an arrow head) or bi-directional (no arrows). The number of objects involved in an association relationship can be designated in various ways:

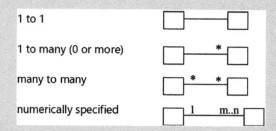

1 to 1	
1 to many (0 or more)	
many to many	
numerically specified	

Figure 6.10   Association relationships.

■ *Aggregation* – a relationship in which one class is made up of other classes (e.g., a Window is made up of four Panels); sometimes call a "whole-part" relationship. Figure 6.11 pictures education class (the whole) as an aggregation of a teacher and a set of students (the parts). In an aggregation relationship, the existence of parts does not depend on the existence of the whole. In another word, the teacher and the students survive after the class is over.

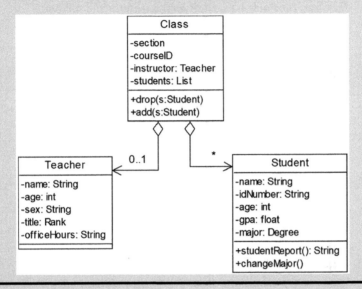

Figure 6.11   Aggregation example.

■ *Composition* – a relationship in which is a type of aggregation where a part object can only belong to one whole object. Figure 6.12 illustrates an Airplane composition. In the composition relationship, the existence of the parts depends on the existence of the whole. For example, if the traffic light fall into the ground, then all the lights also have fallen to ground.

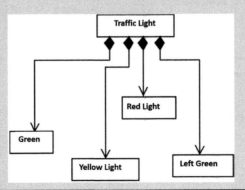

Figure 6.12   Composition example.

■ *Inheritance* – a relationship in which a "child" class is derived from a "parent" class (the child inherits attributes and methods form the parent) (e.g., Temperature Sensor is a subclass of Sensor). Figure 6.13 shows inheritance for reports classes. The TrafficReport and StudentReport classes inherit all of the attributes and methods of the GeneralReport class; but, they may have additional attributes and methods.

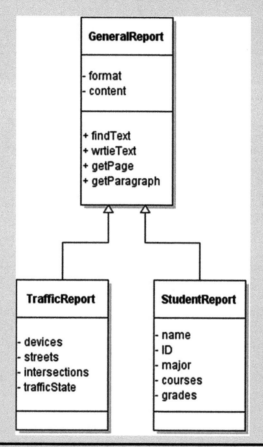

Figure 6.13   Inheritance example.

SEQUENCE DIAGRAMS

A *sequence diagram* is a UML behavioral diagram, which depicts the interaction of the system objects in carrying out a scenario of system execution. The diagram shows the messages exchanged between these objects and the order in which they are exchanged. Figure 6.14 shows an example sequence

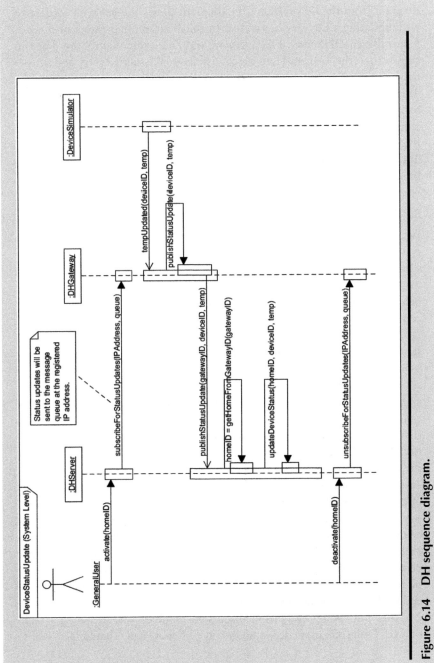

Figure 6.14 DH sequence diagram.

diagram from the DH system. The diagram shows the behavior associated with updating a DH device's status. In the diagram, there are four objects; GeneralUser, DHServer, DHGateway, and DeviceSimulator. The diagram shows that the GeneralUser object initiates this activity of updating a device status by calling the method activate with a particular home ID. This triggers the work of the DHServer object which calls the method subscribeForStatusUpdate. This passing of messages and method invocations continues in order from left to right and top to bottom, until the value of the device is updated by the DHServer (by calling undateDeviceStatus). The vertical dashed line for each object is a *lifeline*, which represents object's life during the interaction.

STATE MACHINE DIAGRAMS

While sequence diagrams are powerful at modeling specific traces of the system, starting with an event stimulating the system and showing how the system moves from one state to the next in response to that stimulus, such diagrams are not intended to provide a comprehensive model of the complete behavior of the system. State machines on the other hand can present the complete behavior of the system or component within the system.

State machine diagrams present the system as a set of finite number of states. A state typically represents the value of system variables at a distinct moment of time. The diagram also shows the allowable transitions between system states. Transitions are triggered by events which are instantaneous and have no duration. In addition to the events, a transition can also be associated with a condition and an action. The condition (if present) indicates that the transition will take place as a response to the event only if the condition is met. An action (if present) represents an activity that takes place once the transition happens, but that does not belong to either of the source or destination states. Figure 6.15 presents state diagram for a traffic light (which might be a subsystem of the TMS discussed earlier). The states of the subsystem are Green, Yellow, and Red lights. The events of elapsed time trigger the changes in state. The condition "business hours" (e.g., 8–10 am and 5–7 pm, each weekday) is a condition that must be true for the transition to take place. Of course, in an actual traffic light subsystem, the diagram would be more complicated, including conditions such "non-business hours", "weekend traffic", "emergency traffic", etc.

After Yao Wang completed discussion of object-oriented design techniques, Disha Chandra led a discussion of how they should proceed to determine a design that would satisfy the DH software requirements. Michel Jackson, the DH software analyst, offered to present a tutorial on how to move from requirements to design.

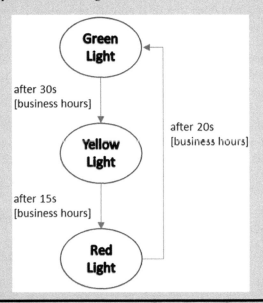

Figure 6.15 **Traffic light state machine diagram.**

MINITUTORIAL 6.6: FROM REQUIREMENTS TO DESIGN. HOW?

One of the most challenging aspects of software development consists of the transition from requirements to design. Once the software requirements have been defined, how does a team begin to translate and map software requirements into design modules? This tutorial provides a brief description of a technique for achieving this transition.

A well-used approach for this transition consists of scanning the requirement specification for "nouns" and "verbs". Nouns might be mapped to actual system components (classes, packages, layers…). For example, in the DH system nouns such as "Sensor", "Alarm", or "Device" might be actual classes. Nouns could also end up being attributes of the components. For example, "temperature" or "humidity" might be attributes of the "thermostat" or "humidistat" device classes. On the other hand, verbs such as "read sensor" or "set alarm status" might end up being operations of the identified components.

It is important to note, that this approach only works if used in iterations, where the initial set of components, attributes, and operations are modified after rigorous analysis. Such analysis is based on maintaining the design principles described in MiniTutorial 6.1: Design Concepts and Principles. It is often the case that components identified in the first cut will lack cohesion or be highly coupled. Repeated analysis and iteration will lead to an improved design.

The rest of the tutorial gives a brief description, in Table 6.2, of a process for developing a high-level object-oriented software design. The process is called "A Simple Object-Based Architectural Design" process (or SOBAD for short).

Table 6.2   SOBAD Process Script

Purpose	• To guide a team through developing and reviewing an object-oriented architectural design for a software development project.
Entry Criteria	• A life cycle software development process • An SRS
Identify Packages	• Divide the problem domain into packages. *Notes:* 1. Designate a name for each package that represents a cohesive part of the problem domain. 2. Use a bi-directional association to show dependencies between the packages.
Identify Classes	• For each package, identify objects (or object classes) from the problem domain. • Describe attributes of each class. • Create a class diagram for the object classes with attributes indicated for each class. *Notes:* 1. You may also add none-domain classes, such as a "controller" class. 2. At this point, you do not need to indicate the data types for the attributes.

(Continued)

Table 6.2 (*Continued*)   SOBAD Process Script

Define Semantics	• Describe the behavior of an object in the class by identifying methods that can be performed on an object; classify each method by type, as either a modifier (changes the state of the object) or an accessor operator (obtains information about the object). • Determine pre-conditions and post-conditions for each method. • Add the method names to the class diagram. • Specify the visibility of each attribute and method (- for private, + for public, and # for protected). *Notes:* 1. Basic "gets" and "sets" for the attributes can be left off the diagram. 2. At this point, you do not need to specify the parameters (and their data types) or return types.
Establish Dependencies	• Determine the relationships between the classes (association, aggregation, and inheritance) within the packages. • Add connectors to the class diagram to show relationships between the classes. *Notes:* If you are not sure about a relationship, use a bi-directional association.
Design External Interfaces	• Design the external interfaces with the software: screen layout for the user interface, report formats, file formats, etc.
Develop Operational Scenarios (Use Cases)	• Develop/enhance operational scenarios for the program that include both normal operation and exception handling. • Create new OTS (Operational Test Scenario) forms for new functionality (if you identified something that was missing from your SRS). Be sure to change the SRS. *Notes:* This may have already been completed during the requirements analysis phase.
Document the Design	• Prepare a high-level document (SDS - Software Design Specification), which includes the package diagram, and the class diagrams, specifying attribute and method properties, and external interface details.

(Continued)

Table 6.2 (*Continued*)   SOBAD Process Script

Evaluate Design	• Use operational scenarios to check the operational behavior of the system and verify that the architectural design includes the required functionality.
	• Evaluate the external interfaces (screen, report, and file layouts)
	• Examine the object/class descriptions, diagrams, and interfaces to insure the following objectives are satisfied:
	  • Object coupling is minimal
	  • Assess whether each object is dependent on too many other objects.
	  • Object cohesion is maximized
	  • Assess whether each object does not present multiple diverse data abstractions: the attributes and operations for a class are all related to the class of objects being defined.
	  • Modifier and accessor operators are identified
	  • Check that each method is either a modifier or an accessor operator.
	  • Proper pre-conditions and post-conditions are included
	  • Ensure that appropriate pre-conditions and post-conditions are defined for each method.
	  • Check that appropriate accessor operators exist for checking for pre-condition and exceptional conditions.
	• Review the class diagram to insure all requirements are included in the design. Each requirement should be traced to one or more classes (and its behavior).

Revise Design	• Based on evaluation, revise the architectural design. If necessary, repeat the design process.
Document Design	• Specify the architectural design in the SDS using the SDS outline [IEEE 1016]. • Revise the System Test Plan, adding modified and new OTSs.
Exit Criteria	• An Architectural Design • A revised System Test Plan

This process is called "simple" because it does not include elements and artifacts that would be needed for a comprehensive design of a complex system. This process is intended for use on a small to moderate size software system that can be developed by small team (3–10 members), over a short period (2–4 months). The same process can also be used by teams following agile process such as Scrum (discussed in Chapter 10). The process does not require prior knowledge or experience with architectural design.

The process is called "object-based" rather than "object-oriented" because it does not include all the elements that might be addressed in an object-oriented design (such as advanced structural and behavioral models). Many of the ideas are based on methods developed by Grady Booch [Booch 1999, 2004] and others.

EXERCISE 6.4: DEVELOPING A DH ARCHITECTURE

The DH Team has made the decision to go with object-oriented design and is now ready to come up with a software architecture for the system. The team will develop a high-level design of the system by defining system components as well as attributes and behavior of these components. The exercise, described at the end of this chapter, deals with how the team will begin the transition between requirements and design by identifying high-level components of a portion of the system.

Design Verification

As the team was completing its initial draft of the SDS, Yao Wang, the System Architect, led the team in discussion of the procedure they would use in verifying the SDS. He reminded the team that the term "verification" referred to the process of evaluating a system or component to determine whether the products of a given development phase satisfy the conditions imposed at the start of that phase (e.g., the design documents cover all the requirements in the SRS). He delivered a short tutorial on design verification. *Note*: Chapter 3 contains additional material validation and verification activities performed throughout the development life cycle.

MINITUTORIAL 6.7: ASSURING DESIGN QUALITY

Given the importance of a system design, it is important that software development teams to be able to determine whether the design properly incorporated requirements specified in the SDS, and to assess the degree to which their design incorporates sound design principles.

The SOBAD Process Script describes a set of activities for evaluating the quality of the design, including the following:

- Use operational scenarios to check the operational behavior of the system and verify that the architectural design includes the required functionality.
- Evaluate the external interfaces (screen, report, and file layouts).
- Examine the object/class descriptions, diagrams, and interfaces to insure that all requirements are included in the design.

Such evaluation would be best handled by a formal design review, in which the developers and other stakeholder review the requirements to ensure their quality; and the reviewers identify and documents design defects.

While inspection and analysis of the design go a long way in determining the strength and weaknesses of a design, there also should be a set of quantitative measures or metrics that measure the quality of the design. This is motivated by the following quote from Tom DeMarco: "you can neither predict nor control what you cannot measure" [DeMarco 1982].

This tutorial provides a description of a suite of metrics used in assessing the quality of an object-oriented design. This set of metrics was developed by Chidamber and Kemerer (CK) [Chidamber 1994]. The Chidamber & Kemerer metrics suite originally consisted of six metrics calculated for each class within an objected-oriented design. In this tutorial, we provide description of four of these metrics; (1) Weighted Methods Per Class (WMC), (2) Coupling Between Object classes (CBO), (3) Response For a Class (RFC), and (4) Lack of Cohesion of Methods (LCOM).

WEIGHTED METHODS PER CLASS (WMC)

The WMC metric measures the complexity of a class by adding the complexity of each of the methods within the class. The complexity of methods is measured using McCabe's Cyclomatic Complexity [McCabe 1976]. The *cyclomatic complexity* of a section of source code can be determined from its control flow graph. The *control flow graph* of a program is a directed graph containing the basic blocks of the program, with an edge between two basic blocks if control may pass from the first to the second. Figure 6.16 contains a flow graph for a program or a method. In this figure, the two circles represent entry to and exit from the module, the rectangles represents a sequence of statements, and the diamonds represent a decision/condition statement. It shows the computation of M, McCabe's cyclomatic complexity.

A class with many complex methods will have a potentially high number of defects and is harder to understand, maintain, and reuse.

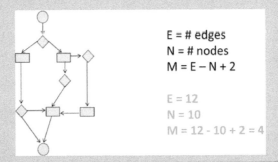

$E = \text{\# edges}$
$N = \text{\# nodes}$
$M = E - N + 2$

$E = 12$
$N = 10$
$M = 12 - 10 + 2 = 4$

Figure 6.16   Cyclomatic complexity.

COUPLING BETWEEN OBJECT CLASSES (CBO)

CBO is a count of the number of non-inheritance classes to which a class is coupled. Two classes are coupled when methods declared in one class use methods or instance variables defined by the other class. For example, the CBO for theTrafficManager class in Figure 6.2 is three. The uses relationship can go either way: both uses and used-by relationships are taken into account; but, only once. Designs where components are highly coupled are typically less efficient and rarely reusable.

RESPONSE FOR A CLASS (RFC)

RFC measures the cardinality of the set of all methods that can be invoked in response to a message to an object of the class or by some method in the class. For example, a method that reads environmental sensors (temperature sensors, light sensors, humidity sensors, power sensors, contact sensors, water sensors, etc.) and saves in the DH database would have a high RFC. If a large number of methods can be invoked in response to a message, the testing and debugging of the class becomes complicated, since it requires a greater level of understanding on the part of the tester. This metric evaluates understandability, testability, and maintainability of the design.

LACK OF COHESION OF METHODS (LCOM)

LCOM measures the degree of similarity of methods by data input variables or attributes (structural properties of classes). LCOM of a class is computed as follows:

1. For each variable within a class, calculate the percentage of the methods that use that variable.
2. Average the percentages calculated in (1) for all variables.
3. Subtract the result in (2) from 100%.

It is important to note that a high resulting number in this calculation will result in a lowly cohesive component or class, which is not desirable. A lowly cohesive class is not a reusable one.

EXERCISE 6.5: MEASURING DESIGN QUALITY

As a process of ensuring good design of components developed by the DH development team, the team uses the CK metrics to evaluate their designed components. The exercise, described at the end of this chapter, deals with how the team measures the quality of its design.

Software Reuse and Design Patterns

In April 202X Jose Ortiz, the Director, *DigitalHomeOwner* Division of HomeOwner Inc, met with the DH development team to emphasize a desire within management to improve on the organization's meeting of its deadlines and schedules. In the meeting, Jose asked the team to suggest ideas to ensure that delivery and deployment of systems for potential clients is done in a more rapid manner.

After a session of brainstorming, Disha Chandra suggested that the team employ more of software reuse. She highlighted the need for reuse of already developed artifacts at all phases of development. She also asked that the team suggest ways not to only reuse previously developed components, but as importantly, to develop artifacts with the vision of using them in future systems.

To achieve these goals, the team recognized that they must put more focus on designing systems for change, and at the same time making use of already established reuse methodologies such as architecture views, styles, and design and code patterns.

MINITUTORIAL 6.8 DESIGN PATTERNS

A *design pattern* is a general reusable solution to a commonly occurring design problem, within a context. A pattern is not a complete design that can be translated into code. It is rather a template that provides guidance for the way to address a problem that tends to happen over and over. Design patterns are different from architecture styles in the sense that architectural styles describe a much high level of abstraction and tend to describe the system at the architectural level, while design patterns have a close relationship with the implementation code.

A well-used set of designed patterns have been introduced in the book *Design Patterns: Elements of Reusable Object-Oriented Software* [Gamma 1995], which describes 23 classic software design patterns. The work (the authors sometimes called the "Gang of Four") breaks the design patterns by type into three categories as follows:

- *Creational* – These are patterns used for creating objects and relieving the programmer from instantiating objects directly.
- *Structural* – These patterns focus on the composition of classes and objects. They also show how objects can be composed to derive new functionality.
- *Behavioral* – These patterns focus on providing effective ways for the interaction between objects.

A description of each of the patterns is provided in terms of the following sections:

- *Intent* – describes the goal of the pattern
- *Motivation* – provides a description of the problem to be solved
- *Applicability* – gives a clear description of the situations where the pattern should be used
- *Structure* – a UML class diagram of the structure of the classes suggested by the pattern
- *Participants* – a description of the classes making up the structure
- *Collaborations* – describes the way objects collaborate with each other
- *Consequences* – provides the advantages and disadvantages of using the pattern
- *Implementation* – gives hints and techniques on implementing the pattern
- *Sample Code* – An illustration of how the pattern can be used in a programming language
- *Known Uses* – gives example of systems that use the pattern from the literature
- *Related Patterns* – gives a list of patterns that are typically used with the pattern described.

Providing a description of design patterns in terms of the categories described above facilitates the use of these patterns as it gives a complete description of what, when, and how to use the specific pattern. Since the [Gamma 1995] book, there have been many new patterns introduced.

Table 6.3 provides a brief description of each of the Gang of four patterns.

EXERCISE 6.6: APPLYING DESIGN PATTERNS

In November 202X, as part of the DH Team's effort to ensure enhanced reusability of developed artifacts, Georgia Magee and Massood Zewail hold an informal discussion to identify the set of possible design patterns to use within their respective assigned development components. To gain a better understanding of design patterns and their applicability, Georgia and Massood discussed a couple of the well-known design patterns and tried to identify how and where they might be used within the different DH design components. The exercise, described at the end of this chapter, deals with how the team work to identify candidate design patterns to use within their design.

Table 6.3 Design Patterns

Creational Patterns	
Pattern	*Description*
Abstract Factory	Provide an interface for creating *families* of related or dependent objects without specifying their concrete classes.
Builder	Separate the construction of a complex object from its representation allowing the same construction process to create various representations.
Factory Method	Define an interface for creating a *single* object, but let subclasses decide which class to instantiate. Factory Method lets a class defer instantiation to subclasses.
Prototype	Specify the kinds of objects to create using a prototypical instance, and create new objects by copying this prototype.
Singleton	Ensure a class has only one instance, and provide a global point of access to it.
Structural Patterns	
Adapter	Convert the interface of a class into another interface clients expect. An adapter lets classes work together that could not otherwise because of incompatible interfaces.
Bridge	Decouple an abstraction from its implementation allowing the two to vary independently.
Composite	Compose objects into tree structures to represent part-whole hierarchies. Composite lets clients treat individual objects and compositions of objects uniformly.
Decorator	Attach additional responsibilities to an object dynamically keeping the same interface. Decorators provide a flexible alternative to subclassing for extending functionality.
Façade	Provide a unified interface to a set of interfaces in a subsystem. Facade defines a higher-level interface that makes the subsystem easier to use
Flyweight	Use sharing to support large numbers of similar objects efficiently.
Proxy	Provide a surrogate or placeholder for another object to control access to it.

(Continued)

Table 6.3 (*Continued*)   Creational Patterns

Pattern	Description
Behavioral Patterns	
Chain of Responsibility	Avoid coupling the sender of a request to its receiver by giving more than one object a chance to handle the request. Chain the receiving objects and pass the request along the chain until an object handles it.
Command	Encapsulate a request as an object, thereby letting you parameterize clients with different requests, queue or log requests, and support undoable operations.
Interpreter	Given a language, define a representation for its grammar along with an interpreter that uses the representation to interpret sentences in the language.
Iterator	Provide a way to access the elements of an aggregate object sequentially without exposing its underlying representation.
Mediator	Define an object that encapsulates how a set of objects interact. Mediator promotes loose coupling by keeping objects from referring to each other explicitly, and it lets you vary their interaction independently.
Memento	Without violating encapsulation, capture and externalize an object's internal state allowing the object to be restored to this state later.
Observer	Define a one-to-many dependency between objects where a state change in one object results in all its dependents being notified and updated automatically.
State	Allow an object to alter its behavior when its internal state changes. The object will appear to change its class.
Strategy	Define a family of algorithms, encapsulate each one, and make them interchangeable. Strategy lets the algorithm vary independently from clients that use it.
Template method	Define the skeleton of an algorithm in an operation, deferring some steps to subclasses. Template method lets subclasses redefine certain steps of an algorithm without changing the algorithm's structure.
Visitor	Represent an operation to be performed on the elements of an object structure. Visitor lets you define a new operation without changing the classes of the elements on which it operates.

Documenting Software Design

In early November 202X, the DH development Team Leader Disha Chandra sends an e-mail to the rest of the DH Team emphasizing the importance of documentation for the DH software development artifacts. In the e-mail, Disha thanked the team for doing a great job in documenting the system requirements in the SRS document. She made the point that the same level of rigor must be used when producing the rest of the artifacts. Disha stressed the importance of such documentation, especially when it comes to the architecture. She noted that she planned to have a meeting to discuss how IEEE Std 1016™-2009, *IEEE Standard for Information Technology—Systems Design—Software Design Descriptions*, and the *Software Engineering Body of Knowledge* (SWEBOK) [Bourque 2014] should influence the documentation of the DH design.

MINITUTORIAL 6.9: DESIGN DOCUMENTATION

As is the case with many of the artifacts associated with software development, the Institution of Electrical and Electronic Engineers (IEEE) provides a standard for documenting the design of software systems: *IEEE Standard for Information Technology—Systems Design—Software Design Descriptions* [IEEE 1016].

The document is organized into multiple viewpoints as follows:

- Context viewpoint
- Composition viewpoint
- Logical viewpoint
- Dependency viewpoint
- Information viewpoint
- Patterns use viewpoint
- Interface viewpoint
- Structure viewpoint
- Interaction viewpoint
- State dynamics viewpoint
- Algorithm viewpoint
- Resource viewpoint

The standard also makes recommendations on how to organize and format a Software Design Description (SDD) (which in this chapter we refer to as the SDS – Software Design Specification). Here is the SDD Template the standard provides:

- Frontspiece
  - Date of issue and status
  - Issuing organization
  - Authorship
  - Change history

- Introduction
  - Purpose
  - Scope
  - Context
  - Summary
  - References
  - Glossary
- Body
  - Identified stakeholders and design concerns
  - Design viewpoint 1
  - Design view 1
  - ...
  - Design viewpoint n
  - Design view n
  - Design rationale

The *Software Engineering Body of Knowledge* (SWEBOK) [Bourque 2014] is an excellent resource for organizing the software design effort. Chapter 2 of SWEBOK provides information and advice on the following design areas:

- Software Design Fundamentals
- Key Issues in Software Design
- Software Structure and Architecture
- User Interface Design
- Software Design Quality Analysis and Evaluation
- Software Design Notations
- Software Design Strategies and Methods
- Software Design Tools

The use of standards and other guidance provides a roadmap for the development of software artifacts, and it gives developers the freedom and flexibility to complete sections of the document in parallel and as appropriate information becomes available or as decisions are made. It also makes those developers become better aware of what decisions must be made and documented before the task can be deemed completed.

EXERCISE 6.8: DOCUMENTING A SOFTWARE DESIGN

In early November 202X, the DH development Team Leader Disha Chandra met with the DH Team to discuss how they would organize and document their software design. Specifically, they would decide how the DH SDS would be organized. The exercise, described at the end of this chapter, deals with how the team carries out this task.

Case Study Exercises

EXERCISE 6.1: UNDERSTANDING DESIGN CONCEPTS AND PRINCIPLES

SCENARIO

In early November 202X, after the DH System Architect, Yao Wang, had completed his presentation on Design Concepts and Principles, he opened discussion about this topic with a set of questions.

LEARNING OBJECTIVES

Upon completion of this module, students will have increased ability to:

- Describe and discuss design concepts and principles.
- Explain whether a design does or does not embody a design concept or principle.
- Incorporate appropriate design principles in design of the DH system.

EXERCISE DESCRIPTION

1. This exercise follows a lecture on design concepts and principles that were presented in MiniTutorial 6.1.
2. As preparation for this exercise students are required to read the following:
   - DH Beginning Scenario
   - *DigitalHome* Need Statement
3. Students are grouped into pairs.
4. Each group discusses and answers the below questions relative to the comments and principles presented in MiniTutorial 6.1:
   a. In Figure 6.1, for the TMS, which module is the least cohesive? Why?
   b. In Figure 6.1, for the TMS, the system features are organized into three separate modules (packages). Why not simply place all features (objects) into a single module (package)?
   c. In the Figure 6.1 TMS modular design considers the *signalState* data element in the *TrafficSignal* class. The data structure for *signalState* is not available to entities external to the TrafficSignal class. What design principle does this represent? Why is it used?
   d. Consider Figure 6.2 for the TrafficManagement package. What is your assessment of its use of appropriate software design concepts and principles (abstraction, encapsulation, information hiding, and the type and degree of coupling and cohesion)?
   e. Are there any potential security problems in either the TMS?
5. The group presents its findings to the rest of the class.

EXERCISE 6.2: SELECTING APPROPRIATE SET OF VIEWS FOR THE DH SYSTEM

Scenario

In early November 202X, the DH development team agreed to come up with the set of architectural views appropriate for arriving at a system architecture for the DH system, which would ensure the satisfaction of the needs of the different stakeholders of the system. After reviewing the need statement and the SRS document the team proposes a set of architectural views to be used.

Learning Objectives

Upon completion of this module, students will have increased ability to:

- Describe various architectural views.
- Describe the similarities and differences between architectural views.
- Determine an appropriate set of views for the software architecture of the DH system.

Exercise Description

1. This exercise follows a lecture on architectural views in MiniTutorial 6.4.
2. As preparation for this exercise students are required to read the following:
   - DH Beginning Scenario
   - *DigitalHome* Need Statement
   - DH SRS Document
   - DH Use Case Model
3. Students are grouped in teams of 3–4 members.
4. Each team discusses the different views approach (Kruchten, SEI, and Siemens) described in MiniTutorial 6.4. They assess the advantages and disadvantages of each view approach for the DH system architecture.
5. The team will then decide on which views approach (Kruchten, SEI, or Siemens) would be most appropriate for the DH system and document the reasons for their decision.
6. The team chooses one member of the team to make an oral report to the class on the team's recommendations.

EXERCISE 6.3: SELECTING AN ARCHITECTURAL STYLE

Scenario

It is mid-November 202X, and the DH development team was ready to select an architecture style to be used to define the DH system architecture.

LEARNING OBJECTIVES

Upon completion of this module, students will have increased ability to:

- Describe various architecture styles.
- Describe elements and relations applicable within a style.
- Analyze and determine the appropriate architecture styles for the DH system.

EXERCISE DESCRIPTION

This exercise follows a discussion of architectural styles in MiniTutorial 6.4.
1. As preparation for this exercise students are required to read the following:
   a. DH Beginning Scenario
   b. *DigitalHome* Need Statement
   c. DH SRS Document
   d. DH Use Case Model
2. Students are grouped in teams of 3–4 members.
3. The team discusses the pros and cons of the various architecture styles relative to their use for the DH system.
4. The team decides on which style (Pipe and Filter, Client-Server, Layered, etc.) would be most appropriate for the DH system and documents the reasons for their decision.
5. The team chooses one member of the team to make an oral report to the class on their decisions.

EXERCISE 6.4: DEVELOPING A DH ARCHITECTURE

SCENARIO

In mid-November of 202X, the DH Team was ready to start developing the high-level design of the system. Georgia Magee suggested starting with the DH Planning and Reporting Requirements, 4.6 in the SRS document, and to identify candidate components (classes, packages, or subsystems) that collaborate to deliver the planning and reporting functionality of the system. The team is required to come up with a first cut of the design by examining nouns and verbs within the SRS for possible components and functions, respectively. The team will use well-established design principles in this and subsequent iterations of the design.

LEARNING OBJECTIVES

Upon completion of this module, students will have increased ability to:

- To use am object-oriented design process to come up with a high-level design of a system.
- Map functionality as described in a SRS document to concrete system components.
- Use design principles in breaking a large system into smaller components.

EXERCISE DESCRIPTION

This exercise follows a lecture on OO design principles and the design process described in MiniTutorials 6.5 and 6.6. The students are also required to read SRS 1.3 prior to the exercise, and highlight the list of verbs and nouns in Section 4.6 of the document.

1. Students are grouped in teams of three to four members.
2. Each team examines the list of nouns in the document and discusses whether each of these represents a possible package or class in the system.
3. The team arrives at a list of potential packages and classes and determines the relationships between them.
4. Each team examines the list of verbs in the document and discuss whether each of these represent a possible function of one of the identified classes, and if so defines a method within the class for that functionality.
5. The team finalizes the first cut of the design using identified packages, classes, attributes, and methods.
6. The team examines the design against designed principles such as cohesion and coupling and makes changes as necessary.
7. The team draws package/class diagrams and completes the below table for each class.
8. One member of the team is selected to present their work to the rest of the class.

Use the following template for this exercise:

Class Name:	
Class Description:	
Attributes:	**Description**
Attribute 1	
Attribute 2	
...	
Methods	**Method Description** (accessor/modifier, pre-/ post-conditions)
Method 1	
Method 2	
...	

EXERCISE 6.5: MEASURING DESIGN QUALITY

SCENARIO

In late November of 202X, as part of ensuring the quality of the design, the System Architect Yao Wang requests that the development team present their design and provide evidence of the quality of their designed component as exhibiting sound design principles. He suggests that the team uses the suite of metrics developed by Chidamber and Kemerer in evaluating their design.

LEARNING OBJECTIVES

Upon completion of this module, students will have increased ability to:

- Identify appropriate set of metrics used in measuring the quality of object-oriented designs.
- Evaluate the design of individual components within a particular design using the CK metrics suite.

EXERCISE DESCRIPTION

This exercise follows a lecture on the CK metrics as described in MiniTutorial 6.8.

1. Students are grouped in pairs.
2. Each team examines the DH artifacts provided below.
3. For each of the artifacts, the team determines which metrics can be used to evaluate the artifacts and comes up with a measure using that metric.
4. The team presents its finding to the rest of the class highlighting their own evaluation based on the used metrics.

ARTIFACTS TO BE EVALUATED

1. The control flow diagram for a method in the class *ContactSensor* of the DH system
2. Class diagram of the DH devices component.

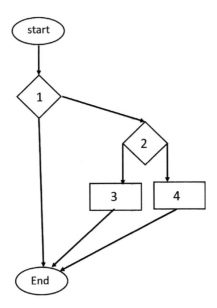

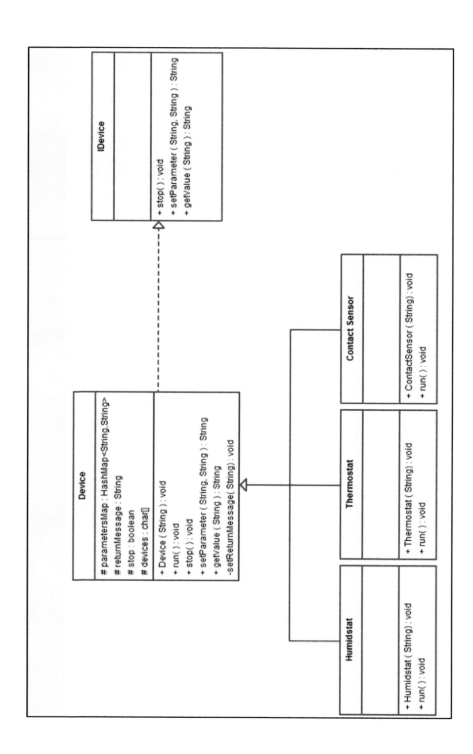

EXERCISE 6.6: APPLYING DESIGN PATTERNS

SCENARIO

In May 202X and as part of the DH Team's effort to ensure enhanced reusability of developed artifacts, Georgia Magee and Massood Zewail held an informal discussion to identify the set of possible design patterns that could be used within DH design components. To gain a better understanding of design patterns and their applicability, Georgia and Massood discussed several of the well-known design patterns and tried to identify their use within the DH components.

LEARNING OBJECTIVES

Upon completion of this module, students will have increased ability to:

- Recognize the value of using design patterns to solve recurring design problems.
- Identify appropriate design patterns for use within a design.

EXERCISE DESCRIPTION

This exercise follows a lecture on design patterns and assigned reading of a set of designed patterns as described in the book [Gamma 1995].

1. Students are assembled into groups of two or three.
2. Each group is responsible to examine the possibility of applying design patterns to a functional feature of the DH system as described the below table.
3. For each functional feature, the team identifies a Gang of Four design pattern which could be used in the design of DH. The team also states its rationale for the selection.
4. One member of the team is selected to present their work to the rest of the class.

Applicable Pattern and Rationale	DH Functional Feature
	Use this pattern to create appliances, sensors, and controllers. Depending on the system's implementation (either simulated or with actual hardware), the abstract factories will be responsible for building components that provide the adequate functionality in relation to the runtime. For example, if the system is running a simulation the "SensorFactory" will be responsible for building virtual sensors that can simulation real-life temperature sensors, humidity sensors, etc.

(Continued)

Applicable Pattern and Rationale	DH Functional Feature
	Use this pattern to create the home layout. The home layout is composed of home compartments, which in turn, are composed of home components. The home compartments will provide an abstraction to define rooms, corridors, garage(s), bathrooms, and so on, while the home components will be used to define appliances, controllers, and sensors.
	Use this pattern to define commands that change the behavior of an appliance or device via a controller. For example, the thermostat controller will use the air-conditioner-on and air-conditioner-off commands to control the air conditioner.
	Use this pattern to add functionality to a user object dynamically. For example, in DH system the General user is able to view readings and modify settings on various devices. The Master user, in addition to being able to perform the same functions as the General user, is also able to create and manage users' accounts. A Technician is able to have the same functionality as the Master user, which means he or she is able to perform the same tasks as the General and Master users.

EXERCISE 6.8: DOCUMENTING A SOFTWARE DESIGN

Scenario

In early September 202X, the DH development Team Leader Disha Chandra met with the DH Team to discuss how they would organize and document their software design. Specifically, they would decide the how the DH SDS would be organized.

LEARNING OBJECTIVES

Upon completion of this module, students will have increased ability to:

- Describe and understand the contents of software design standards and guidelines.
- Analyze and determine an appropriate way to organize and document a software system design, with special emphasis on software architecture.

EXERCISE DESCRIPTION

This exercise follows a discussion of design documentation in MiniTutorial.

1. As preparation for this exercise students are required to read the following:
   a. IEEE Standard for Information Technology—Systems Design—Software Design Descriptions [IEEE 1016]
   b. Software Engineering Body of Knowledge (SWEBOK) [Bourque 2014]
   c. DH SRS Document
2. Students are grouped in teams of 3–4 members.
3. The team discusses the contents of the IEEE Standard and the SWEBOK, and what parts are most useful and applicable to the DH SDS.
4. The team decides on the content and the organization of the DH SDS.
5. The team chooses one member of the team to make an oral report to the class on their decisions.

Chapter 7

Constructing *DigitalHome*

Note: The Preface includes the statement "It is assumed the audience has some basic understanding of the nature of computing, and can use fundamental program constructs (sequence, selection, and repetition) and write simple programs in a high-level object-oriented programming language (such as Java)". So, in this Chapter, basic programming concepts and techniques are not covered, and certain construction principles are illustrated using the Java programming language.

Build/Integration Plan

In Chapter 2, the *DigitalHome* Development Process was introduced (now reproduced in Table 7.1). The process includes the development of a *Build/Integration Plan*. The plan would detail an incremental development approach, which would include the activities specified in the "Construction Increment 1, 2,..." section. These construction activities make up what is often referred to as *Software Construction*, as described in Chapter 3 of the SWEBOK [Bourque 2014]: "The term software construction refers to the detailed creation of working software through a combination of coding, verification, unit testing, integration testing, and debugging".

Table 7.1   *DigitalHome* Development Process Script

Purpose	• Guide the development of the *DigitalHome* software system
Entry Criteria	• Identification of a problem or opportunity, which may need a software solution
Phase	*Activity*
Project Inception	• Establish the business case and feasibility for the project. • Determine the scope of the project. • Develop a customer Need Statement. • Develop a High-Level Requirements Document
Project Launch	• Establish development team, and their roles, responsibilities, and goals. • Analyze Need Statement • Establish a Development Process • Identify External Interfaces and create a Context Diagram • Create a Conceptual Design • Determine a Development Strategy • Research and study of possible project technology
Planning	• Develop Task, Schedule, and Resource Plans • Develop a Risk Plan • Develop a Quality Plan • Develop a Configuration Management (CM) Plan • Determine project standards and tools • Perform project planning/tracking

(Continued)

Table 7.1 (*Continued*)   *DigitalHome* Development Process Script

Purpose	• Guide the development of the *DigitalHome* software system
Entry Criteria	• Identification of a problem or opportunity, which may need a software solution
Phase	*Activity*
Analysis	• Review the Need Statement and prepare for requirements elicitation • Elicit requirements from potential customers and users • Analyze requirements and build object-oriented analysis models • Specify and document requirements in a Software Requirements Specification (SRS) • Inspect and revise the SRS • Obtain customer agreement to requirements • Develop a System Test Plan • Develop a Requirements Traceability Matrix (RTM) • Develop a User Manual • Track, monitor, and update plans (Task/Schedule/Quality/CM/Risk) & RTM
Architectural Design	• Develop an Object-Oriented (OO) high-level design model • Refine the OO analysis model • Develop additional design models • Specify and document design in an Software Design Specification (SDS) • Develop a Build/Integration Plan • Review and revise the SDS • Consult with customer, and revise SRS and SDS, as appropriate • Track, monitor, and update plans (Task/Schedule/Quality/CM/Risk) & RTM
Construction Increment 1, 2,...	• Complete Class Specifications (data types and operation logic) • Develop incremental Integration Test Plan (Class/Integration) • Set up an appropriate test environment (e. g., test harness, stubs/drivers)

(Continued)

Table 7.1 (*Continued*) *DigitalHome* Development Process Script

Purpose	• Guide the development of the *DigitalHome* software system
Entry Criteria	• Identification of a problem or opportunity, which may need a software solution
Phase	*Activity*
Construction Increment 1, 2,...	• Review the Class Specifications • Write source code • Review and test the source code • Integrate code with previous increments • Perform Increment (Class/Integration) Test • Consult with customer, and revise SRS, SDS, and code, as appropriate • Track, monitor, and update plans (Task/Schedule/Quality/CM/Risk) & RTM
System Test	• Update test environment • Update System Test Plan • Perform the System Test • Consult with customer, and revise requirements, design and code as appropriate • Track, monitor, and update plans (Task/Schedule/Quality/CM/Risk) & RTM
Acceptance Test	• Prepare Acceptance Test materials (including a user's manual) • Set up customer test environment • Customer performs Acceptance Test • Revise SRS, SDS, code, and User Manual as appropriate • Track, monitor, and update plans (Task/Schedule/Quality/CM/Risk) & RTM
Postmortem	• Perform postmortem analysis of project process and product quality
Exit Criteria	• Completed SRS, SDS, RTM, User Manual, System Test Plan • Source code for a thoroughly reviewed and tested program • Completed plan documentation • Project postmortem report

In a team meeting following the specification of the DH Architecture, Disha Chandra asked Massood Zewail to lead the effort to develop a Digital Home Build/ Integration Plan. Massood presented a high-level diagram (see Figure 7.1) and led a discussion of how to break the software construction into increments and develop a Build/Integration Plan.

EXERCISE 7.1: DEVELOPING BUILD/INTEGRATION PLAN

After the team reviewed the Design Diagram in Figure 7.1, the team discussed development of a DH Build/Development Plan. The exercise, described at the end of this chapter, deals with how the DH Team developed the plan.

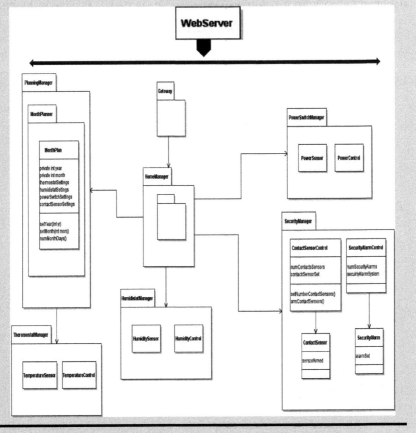

Figure 7.1   DH design diagram.

Construction Fundamentals

After the team completed a DH Build/Development Plan, Disha Chandra said to the team "I know you are all have experience constructing and testing software modules, but I think it would be good for us to review some fundamentals of software construction". Disha asked Georgia Magee to prepare and deliver a tutorial on Software Construction.

**MINITUTORIAL 7.1: SOFTWARE
CONSTRUCTION FUNDAMENTALS**

Georgia began with a discussion of how to construct the software for the example ATM system depicted in Figure 7.2. For example, suppose one wanted to construct the class *BankAccount*. Here are some of the activities, which would need to be carried out:

- Plan for testing and integration of the class with the rest of the ATM system. If a Build/Integration Plan has been developed, this provides the appropriate foundation for this step.
- Complete Class Specifications (data types and method specification) for *Customer, Money, BankAccount*, and the other classes in Figure 7.2. This might include the addition of new fields and methods.

For example, in Figure 7.2 the *Customer* class would need the specification of the data types for the attributes name, address, and SSN. Also, the methods in the class would need to be determined (e.g., a *changeAddress* method).

- Develop the detailed design and write the source code for each method.

A first step might be the specification of pre-conditions and post-conditions (covered later in this Chapter).

The use of pseudocode for detailed design of the method (especially for complicated methods) would describe the method logic without concern about language syntax. (The approach of test-driven design is covered later in this Chapter).

- Review and test the source code for each method. A rigorous review will reduce overall development time and improve system quality. (See Chapter 3 for more detail.)
- Set up an appropriate test environment (e. g., test harness, stubs/drivers). (See Chapter 3 for more detail.)

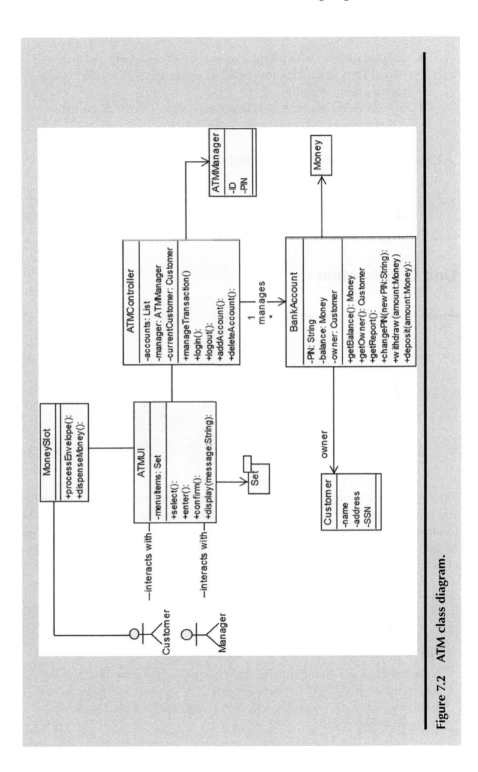

Figure 7.2   ATM class diagram.

- Integrate the code with previous increments of the system. For example, a Build/Integration Plan for the ATM would specify the order of construction and integration.
- Perform an Increment (Class/Integration) Test. (This is discussed later in this chapter.)

EXERCISE 7.2: SPECIFYING A DH CLASS

After Georgia's tutorial on software construction, the team reviewed the Design Diagram in Figure 7.1, the team began a discussion of a more detailed specification of the *MonthPlan* class. The exercise, described at the end of this chapter, deals with how the *MonthPlan* class would be specified with more detail.

Unit Construction

The *DigitalHome* team decided to use an object-oriented methodology for design and construction, with Java as the implementation language. A Java method is the fundamental implementation unit. Team Leader Disha Chandra asked Georgia Magee to continue her tutorial on software construction by addressing issues specifically related to constructions of a Java method.

MINITUTORIAL 7.2: METHOD CONSTRUCTION

PRE-CONDITIONS AND POST-CONDITIONS

A *pre-condition* is a condition that must be true before a method is called. For example, in the ATM system (Figure 7.2), consider the *BankAccount* method *withdraw* (amount: Money). A pre-condition for withdraw would be "amount <= balance".

A *post-condition* is a condition that must be true after a method is executed. Consider the *BankAccount* method *deposit* (amount: Money) would be new balance=old balance+amount.

UNIT TESTING

Unit testing is used to determine defects when executing code for a software unit. A *test plan* for unit testing would consist a set of test cases, where a *test case* tests a single feature of the unit. There is more information about Unit Testing in Chapter 3.

In Chapter 2, we discuss the eXtreme Programming (XP) process [Beck 2000]. One of XP's features is Test-Driven Development (TDD). TDD is a development process for a software unit (e.g., a class method) that consists of several very short development cycles, where each cycle consists of the following steps:

- Create a unit test case, testing a single feature of the unit (not yet in the unit).
- Run the test, which should fail because the program lacks that feature.
- Modify the unit code just to make the test run correctly.
- "refactor" the code (clean it up), so that it is manageable (it may need new variable names, additional documentation, or be divided into multiple units).
- Repeat the above steps and aggregate the test cases in a single test plan.

A similar but less comprehensive approach is to write the entire unit test plan before one begins writing any code. This has the following advantages:

- It motivates the developer to carefully review the requirements for the unit.
- Test planning for possible execution errors early-on prompts the developer to write code for exception handling.
- Since a test plan has to be created at some point, doing this before unit coding is no costlier than waiting until after the code is written.

For example, suppose a developer was writing the code for the *withdraw (amount: Money)* method in Figure 7.2. She could create a test plan like the one in Figure 7.3.

ALGORITHM DESCRIPTION USING PSEUDOCODE

Pseudocode is a notation resembling a simplified programming language, which is used to depict the logic and structure of a program's design. Some of the advantages of using pseudocode are as follows:

Test Plan/Report

Project: ATM: BankAccount   **Date:** 2/7/2023
Unit: withdraw(amount: Money)   **Developer:** William Sutton

Test Description / Data	Expected Results	Test Experience / Actual Results
amount = balance	balance' = 0.00	
amount = 0.4* balance	balance' = 0.6 * balance	
amount = 0	error message	
amount = -1.0	error message	
amount = balance + 1.0	error message	

Figure 7.3   *BankAccount* test plan.

- ■ The program designer can concentrate on the algorithm for the program, and not be distracted by programming style and syntax.
- ■ The pseudocode can be translated in different programming languages without modification.
- ■ The pseudocode can be reviewed for requirements and logic errors.

Figure 7.4 shows a pseudocode for a program than that computes the greatest common divisor for two integers.

Figure 7.5 shows the Java method based on the GCD pseudocode.

```
function GCD (integer a, integer b)
  declare integer temp, value
   begin GCD
  1. a := abs(a);
  2. If (a=0 and b=0) then
  3.      raise exception
  4. else if (a=0) then
  5.      value := b;   // b is the GCD
  6. else
  7.      while (b ≠ 0 ) loop
  8.          temp := b
  9.          b := a mod b
  10.         a := temp
  11.     end loop
  12.     value := a;  // a is the GCD
  13. end if;
  14. return value;
  end GCD
```

Figure 7.4   GCD algorithm.

```
/************************************************************************
 Author: Euclid
 Name: GreatestCommonDivisor
 Version: 1.0 2/16/201X
 Purpose: computes the GCD of two integers
 ************************************************************************/
public class GreatestCommonDivisor {
     public static int gcd(int a, int b)  {
        int temp, value;

        a = Math.abs(a);
        if (a == 0 & b == 0) {
            System.out.println("GCD is only valid for non-
     zero integers");
            System.exit(0);

        } else if (a == 0) {
             value = b;     // b is the GCD
        } else {
            while (b != 0 ) {
                temp = b;
                b = a % b;
                a = temp;
                }
        }
        value = a;  // a is the GCD
        return value;
     }
}
```

Figure 7.5   GCD java code.

CODING STANDARDS

Code documentation is an important element of unit construction. Good documentation, using a coding standard, can have the following advantages:

- Provide the developer with structure and layout, which eases the code developing and organizing efforts.
- Improve the readability of the code by those other than the developer.
- Makes maintenance of the unit easier.
- Improves tracing and correcting execution.

Although there are a number of coding standards for Java, one of the first and most widely used is *Java Code Conventions* [Sun 1997]. It provides information on the following:

- Declaration format for classes, methods, and variables.
- Comment advice and style.

- Statement format and naming conventions.
- Use of indentation, white space, and line length to improve readability.

CODE REVIEW

As we discussed in Chapter 3, software quality is difficult and expensive to achieve when developers depend solely on software testing. Software inspection and reviews are cost-effective ways to detect and correct defects. This includes unit development; unit testing is essential, but review of the unit code will produce a higher quality product and reduce testing time.

Table 7.2 provides a checklist of activities to be performed in a code review of unit code. The checklist is based on ideas presented in [Humphrey 1995].

EXERCISE 7.3: PRE-CONDITIONS AND POST-CONDITIONS OF A DH CLASS

After Georgia's tutorial on Method Construction, the team discussed and determined the pre-conditions and post-conditions for a DH class. The exercise, described at the end of this chapter, deals with how the DH Team decided on these conditions.

EXERCISE 7.4: TEST PLANNING FOR A DH UNIT

After Georgia's tutorial on Method Construction, the team discussed and agreed on their test planning strategy. The exercise, described at the end of this chapter, deals with how the DH Team creates a test plan for a class method.

EXERCISE 7.5: DEVELOPING A DH UNIT

After Georgia's tutorial on Method Construction, the team discussed development of DH class methods. The exercise, described at the end of this chapter, deals with how the DH Team members would develop a class method.

Disha Chandra asked Georgia Magee to continue her tutorial on software construction by addressing issues related to writing code that does have security weaknesses.

Table 7.2 *DigitalHome* Code Review Guide

Purpose	• Guide the review of *DigitalHome* Software Unit Code
Entry Criteria	• Completion of the code for software unit, such as for a class method

Step	*Activities*	*Description*
1	Completeness	• Check the program against the requirements specification (SRS) to ensure that it includes all required functionality. • Check the program against the design specification (SDS) to ensure that all required unit interfaces and functionality are implemented.

2	Checklist	Use the following code review checklist to find all the defects in the program:
		• Verify that the program flow and all function logic are consistent with the detailed design.
		• Trace through all loops and recursive operations.
		  • Ensure that every loop is properly initiated and terminated.
		  • Check that every loop is executed the correct number of times.
		• Check every method call to ensure that it exactly matches the definition for formats and types.
		• Verify that each variable and parameter has exactly one declaration and is only used within its declared scope.
		• Check all variables, arrays, and indexes to ensure that their use does not exceed declared limits.
		• Check all begin-end pairs, including cases where nested ifs could be misinterpreted.
		• Check Boolean conditions.
		• Check every statement for instruction format, spelling, and punctuation. Check that all pointers are properly referenced.

(Continued)

Table 7.2 (*Continued*)   *DigitalHome* **Code Review Guide**

Purpose		• Guide the review of *DigitalHome* Software Unit Code
Entry Criteria		• Completion of the code for software unit, such as for a class method
Step	*Activities*	*Description*
		• Check all input-output formats. • Check that every variable, parameter, and key word is properly spelled. • Ensure that all commenting is accurate and according to standard. • Check that unit structure is comprehensible with appropriate use of indentation and white space.
3	Corrections	Keep a log of the defects found; and following the code review, correct all the defects.
4	Final check	• All the defects found have been corrected. • The corrections have been checked for correctness.

MINITUTORIAL 7.3: SECURE CODING

Secure coding is the practice of developing program code so that it protects against the introduction of security weaknesses into the software. Defects and logic flaws are typically the cause of common software vulnerabilities.

The SEI CERT (Software Engineering Institute Computer Emergency Response Team) partners with government, industry, law enforcement, and academia to improve the security and resilience of computer systems and networks. On the website (https://wiki.sei.cmu.edu/confluence/display/seccode/) there are coding standards for commonly used programming languages such as C, C++, Java, and Perl, and the Android™ platform. The site also discusses the Top 10 Secure Coding Practices, listed in Table 7.3.

A classic example of a a secure coding problem is the *buffer overflow problem*. Buffer overflow occurs when data is input or written beyond the

Table 7.3   Top Ten Secure Coding Practices

Validate input. Validate input from all untrusted data sources.
Heed compiler warnings. Compile code using the highest warning level available for your compiler and eliminate warnings by modifying the code.
Architect and design for security policies. Create a software architecture and design your software to implement and enforce security policies.
Keep it simple. Keep the design as simple and small as possible. Complex designs increase the likelihood that errors will be made in their implementation, configuration, and use.
Default deny. Base access decisions on permission rather than exclusion. This means that, by default, access is denied and the protection scheme identifies conditions under which access is permitted.
Adhere to the principle of least privilege. Every process should execute with the least set of privileges necessary to complete the job. Any elevated permission should only be accessed for the least amount of time required to complete the privileged task.
Sanitize data sent to other systems. Sanitize all data passed to complex subsystems such as command shells, relational databases, and commercial off-the-shelf (COTS) components.
Practice defense in depth. Manage risk with multiple defensive strategies, so that if one layer of defense turns out to be inadequate, another layer of defense can prevent a security flaw from becoming a vulnerability.
Use effective quality assurance techniques. Good quality assurance techniques can be effective.   **Adopt a secure coding standard.** Develop and/or apply a secure coding standard for your target development language and platform.

allocated bounds of an object causing a program crash or creating a vulnerability that attackers might exploit.

EXERCISE 7.6: BUFFER OVERFLOW PROBLEM

After Georgia's tutorial on Secure coding, the team discussed the buffer overflow problem development. The exercise, described at the end of this chapter, deals with how the DH Team members created code that exhibited the buffer overflow problem.

Case Study Exercises

EXERCISE 7.1: DEVELOPING BUILD/INTEGRATION PLAN

SCENARIO

After the team reviewed the Design Diagram in Figure 7.1, the team discussed development of a DH Build/Development Plan. This exercise deals with how the DH Team developed the plan.

LEARNING OBJECTIVES

Upon completion of this module, students will have increased ability to:

- Describe and discuss a Build/Development Plan.
- Separate design and development features into increments.
- Create a plan for design, construction, and integration of a software system.

EXERCISE DESCRIPTION

1. This exercise follows a lecture on the Digital Home Build/Integration Plan.
2. As preparation for this exercise students are required to read the following:
   - DH Beginning Scenario
   - *DigitalHome* Need Statement
   - *DigitalHome* Software Requirements Specification
3. Students are grouped into teams of 3–4.
4. Each team reviews and discusses the Architectural Design and Construction portions of Table 7.1 and Figure 7.1.
5. Each team decides on the number of construction increments and selects which elements of Figure 7.1 will be assigned to each increment.
6. Each team creates a DH Build/Design Plan with a schedule for order of development and integration of each increment (including the estimated duration for construction of each increment).
7. The team presents its findings to the rest of the class.

EXERCISE 7.2: SPECIFYING A DH CLASS

SCENARIO

After the team reviewed the Design Diagram in Figure 7.1, the team began a discussion of a more detailed specification of the *MonthPlan* class.

LEARNING OBJECTIVES

Upon completion of this module, students will have increased ability to:

- Describe and discuss how a Java class should be specified.
- Better understand *DigitalHome* specification and implementation.
- Create a plan for design and development of a software system.

EXERCISE DESCRIPTION

1. As preparation for this exercise students are required to read the following:
   - DH Beginning Scenario
   - *DigitalHome* Need Statement
   - *DigitalHome* Software Requirements Specification (with special attention to Section 4.6)
2. Students are grouped into teams of 3–4.
3. Each team reviews and discusses the Architectural Design and Construction portions of Table 7.1 and Figure 7.1.
4. Each team reviews and discusses the *MonthPlan* class in of Figure 7.1 using the below questions:
   - Why are year and month specified as type int?
   - How should an attribute like thermostatSettings be specified?
   - How should the method setYear() be specified (i.e., the specification of the method header)?
   - What attributes or methods should be added to *MonthPlan*?
5. Each team creates a report about its findings.
6. The team presents its findings to the rest of the class.

EXERCISE 7.3: PRE-CONDITIONS AND POST-CONDITIONS OF A DH CLASS

SCENARIO

After the team reviewed the Design Diagram in Figure 7.1, the team began a discussion of pre-conditions and post-conditions of the class *ContactSensorsControl*.

LEARNING OBJECTIVES

Upon completion of this module, students will have increased ability to:

- Describe and discuss pre-conditions and post-conditions.
- Identify *DigitalHome* pre-conditions and post-conditions.
- Develop error-free methods.

EXERCISE DESCRIPTION

1. As preparation for this exercise students are required to read the following:
   - DH Beginning Scenario
   - *DigitalHome* Need Statement
   - *DigitalHome* Software Requirements Specification (with special attention to Section 4.4)
2. Students are grouped into teams of 3–4.
3. Each team reviews and discusses the ContactSensorsControl class in of Figure 7.1, and identifies appropriate pre-conditions and post-conditions for the methods.
4. Each team creates a report about its findings and presents it to the rest of the class.

EXERCISE 7.4: TEST PLANNING FOR A DH UNIT

SCENARIO

After the team reviewed the Design Diagram in Figure 7.1, they discussed and agreed on their test planning strategy. This exercise deals with how each DH Team member would create a test plan for the *numMonthDays* method in the *MonthPlan* class. The following is a brief specification for the method:

```
/************************************************************************

   Name: numMonthDays

   Purpose: sets the values of the Year and Month fields and returns the

        number of days in that Year and Month                          *

   *********************************************** ***********************/

   public static int numMonthDays() throws IOException {

      Note: the setMonth and setYear methods zin MonthPLan are used in this
method.
```

LEARNING OBJECTIVES

Upon completion of this module, students will have increased ability to:

- Plan tests for units of code.
- Develop error-free methods.
- Understand *DigitalHome* functionality.

EXERCISE DESCRIPTION

1. As preparation for this exercise students are required to read the following:
   - DH Beginning Scenario
   - *DigitalHome* Need Statement
   - *DigitalHome* Software Requirements Specification (with special attention to Section 4.6)
2. Teams of 3–4 are formed and members work individually on a test plan.
3. Each team reviews and discusses the numMonthDays method in the MonthPlan class, in Figure 7.1, and individuals develop test plans for the method.
4. When test plans are completed, the team compares and contrasts their plans.
5. The team reports on their test planning to the rest of the class.

EXERCISE 7.5: DEVELOPING A DH UNIT

SCENARIO

After the team reviewed the Design Diagram in Figure 7.1, they discussed class method development. This exercise deals with how each DH Team member would develop the *numMonthDays* method in the *MonthPlan* class. The following is a brief specification for the method:

```
/***************************************************************
   Name: numMonthDays
   Purpose: sets the values of the Year and Month fields and returns the
         number of days in that Year and Month                        *
   ********************************************************* ****************/
   public static int numMonthDays() throws IOException {
```
Note: the setMonth and setYear methods in MonthPLan can be used in this method.

LEARNING OBJECTIVES

Upon completion of this module, students will have increased ability to:

- Develop units of code.
- Develop error-free methods.
- Understand *DigitalHome* functionality.
- Perform a code review of a unit of code.

EXERCISE DESCRIPTION

1. As preparation for this exercise students are required to read the following:
   - DH Beginning Scenario
   - *DigitalHome* Need Statement
   - *DigitalHome* Software Requirements Specification (with special attention to Section 4.6)
2. Teams of 3–4 are formed and members work individually on method development.
3. Each team reviews and discusses the *numMonthDays* method in the *MonthPlan* class, in Figure 7.1, and individuals develop the following:
   - A pseudocode algorithm for the *numMonthDays* method.
   - The source code for the numMonthDays method. Use Java or some other object-oriented programming language.
4. When code reviews are completed, the team compares and contrasts their reviews.
5. The team reports on their code development activities to the rest of the class.

EXERCISE 7.6: BUFFER OVERFLOW PROBLEM

SCENARIO

After Georgia's tutorial on Secure coding, the team discussed the buffer overflow problem. Each DH Team member was asked to create code that exhibited the buffer overflow problem.

LEARNING OBJECTIVES

Upon completion of this module, students will have increased ability to:

- Understand the buffer overflow problem.
- Develop error-free code.

- Understand the importance of secure coding.
- Perform a code review of a unit of code.

EXERCISE DESCRIPTION

1. Teams of 3–4 are formed and members work individually on developing an example method that has a buffer overflow problem.
2. Each team reviews and discusses the examples of buffer overflow code.
3. The team reports on their code development activities to the rest of the class.

Chapter 8

Maintaining *DigitalHome*

Maintenance Fundamentals

On 11/15/202X, the DH development team had a meeting to discuss their progress in developing the *DigitalHome* architecture. In the meeting, Michel Jackson raised a tangential issue:

> It seems we are making good progress in developing the requirements and design documents, but I would like to raise an issue where I have seen too many projects give it little attention or delay too long thinking about it. Early consideration of the role of software maintenance in our work can help produce high quality long-lived systems.

The team discussed this issue, and some admitted they had little knowledge or experience with software maintenance. Disha Chandra asked Michel if he would prepare and deliver a set of tutorials on software maintenance.

MINITUTORIAL 8.1: SOFTWARE MAINTENANCE FUNDAMENTALS

Michel stated that the following references were used in the development of the maintenance tutorials:

- Chapter 5 of *Software Engineering Body of Knowledge* (SWEBOK) [Bourque 2014].
- IEEE Std 14764-2006, Software Engineering-System Life Cycle Processes-Maintenance [IEEE 14764].

The IEEE Standard *Systems and Software Engineering Vocabulary* [IEEE 24765] defines *software maintenance* as "the process of modifying a software system or component after delivery to correct faults, improve performance or other attributes, or adapt to a changed environment". Software maintenance is part of *software evolution*: the term used to refer to the process of developing software initially, then repeatedly changing it over time to adapt to new requirements or adapt to new environments. The SWEBOK states "The objective of software maintenance is to modify existing software while preserving its integrity". It does this in the following ways:

- Correcting errors in the software.
- Adapting software to a new environment.
- Enhancing functionality or improving performance.

Software maintenance is different than hardware maintenance – software does not wear out. However, software maintenance can be extremely expensive [Dehaghani 2013]:

- Maintenance costs can account for over 60% of the total software life cycle costs.
- About 20% of maintenance costs are spent on correcting errors.
- Half of the maintenance costs can be consumed in trying to understand the software.

TYPES OF MAINTENANCE

Maintenance is grouped into the following categories:

- *Corrective Maintenance* – changes to correct defects discovered after delivery.
- *Adaptive Maintenance* – modification of a software product performed after delivery to keep a software product usable in a changed

or changing environment (e.g., a new operating system or a different hardware platform).

■ *Perfective Maintenance* – changes requested/needed for enhancement of the software after delivery.

■ *Preventive Maintenance* – changes made to software to prevent failures or improve its maintainability.

In Chapters 4–6, we discussed a Traffic Management System (TMS) – a system that manages and controls the traffic in a medium size city (e.g., 100,000 population). Figure 8.1 depicts a high-level diagram for the TMS design.

Consider the following situations that would require maintenance of the TMS:

Corrective Maintenance Situation

■ A requirement for the TMS specified that the traffic signals must change within less than one second. However, when the system is put into operation, it is discovered that signal changes typically take more than one second.

■ The method *getClockTime()* specified that the time be returned in military format (e.g. 2130 hours for 11:30 pm). However, the method was implemented using am/pm time (e.g., time was returned as 11:30 pm, rather than as 2130).

Adaptive Maintenance Situation

■ The client (the city government) for the TMS decided to make a change to the signal lights, so that they have sensors that can detect when there is no traffic in a lane. The client wants to add a requirement that would specify the signal light would not change when there is no traffic sensed in a traffic lane.

Perfective Maintenance Situation

■ A maintenance team member experiences a problem in changing TMS code because of poor documentation. She recommends that a code documentation standard be adopted and be applied to improve TMS code readability.

Preventive Maintenance Situation

■ Because of numerous faults caused by software design defects, the maintenance team decides to conduct an inspection of the design documents and fix any problems found.

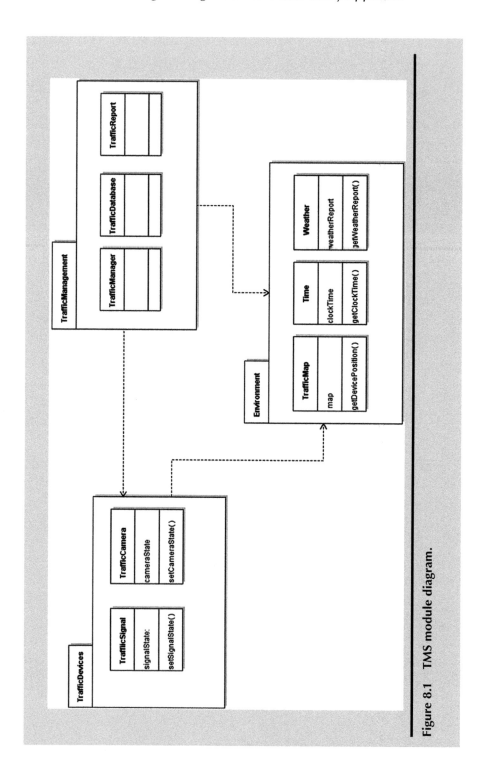

Figure 8.1 TMS module diagram.

MAINTENANCE COSTS

As we mentioned earlier, maintenance costs are a significant part of the lifetime costs of a software system. In Figure 8.2 [Lientz 1981], there is a 1980 depiction of the average distribution of maintenance time of software systems. Unfortunately, the high percentage of corrective effort at 21% has not changed much in the last 40 years.

The estimation of the cost of software maintenance is an important part of planning for software maintenance. In Chapter 4, we discussed methods for estimating the cost of the initial development of a software product. The discussion covered techniques such as Expert Judgment, the Delphi method, and Statistical/Parametric Methods. The standard *Software Engineering-System Life Cycle Processes-Maintenance* [IEEE 14764] states that "the two most popular approaches to estimating resources for software maintenance are the use of parametric models and the use of experience".

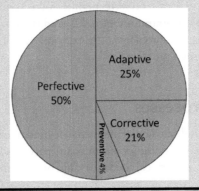

Figure 8.2   Maintenance time.

MAINTENANCE MEASURES

Two essential elements affecting the value of software maintenance are the time to fix defects and the quality of the fixes [Kan 2002]. Measurement of these values and other related ones can assist in estimating software maintenance costs and can improve the quality of software changes.

A key measurement related to the fix backlog (defects discovered, but not yet fixed) is the *backlog management index* (BMI):

$$\mathrm{BMI} = \frac{\text{Number of defects fixed during the month}}{\text{Number of defects discovered during the month}} \times 100\%$$

If BMI is over 100% then the backlog is reduced; if it less than 100% the backlog is increased.

Other measures include the following [Bourque 2014]:

- Software size, complexity, and understandability.
- *Analyzability* – measures of the resources expended in trying either to diagnose deficiencies or causes of failure or to identify parts to be modified.
- *Changeability* – measures of the maintainer's effort associated with implementing a specified modification.
- *Stability* – measures of the unexpected behavior of software, including that encountered during testing.
- *Testability* – measures of the effort in trying to test the modified software.

EXERCISE 8.1: IDENTIFYING POTENTIAL DH MAINTENANCE PROBLEMS

After Michel's tutorial on software maintenance fundamentals, the team reviewed the Design Diagram in Figure 8.1; and the team began a discussion of potential maintenance problems. The exercise, described at the end of this chapter, deals with such problems.

Maintenance Processes

After Michel finished his discussion of maintenance fundamentals, he emphasized that since software maintenance occupies a significant part of software system's lifespan (both in cost and time), it is critical to identify and organize maintenance activities.

MINITUTORIAL 8.2:   SOFTWARE MAINTENANCE PROCESSES

As described in [IEEE 14764], a *maintenance process* should include the following activities (see Figure 8.3):

- *Process Implementation* – the maintenance team establishes the plans and procedures which are to be executed during the maintenance process.
- *Problem and Modification Analysis* – the maintenance team does the following:
  - analyzes Modification Requests (MRs) or Problem Reports (PRs)
  - replicates or verifies the problem
  - develops options for implementing the modification
  - documents the MR/PR, the results, and execution options
  - obtains approval for the selected modification option

- *Modification Implementation* – the maintenance team develops and tests the modification of the software product.
- *Maintenance Review/Acceptance* – this activity ensures that the modifications to the system are correct and that they were accomplished in accordance with approved standards and adhere to current requirements and design documents.
- *Migration* – if a system must be modified to run in a different environment (a "migration"), the maintenance team needs to determine the actions/steps required to migrate the system.
- *Retirement* – once a software product has reached the end of its useful life, it must be retired; and a retirements analysis should be performed that includes things such as the following:
  - Determine if the retired system should be replaced with a new software product.
  - Access what should be done with data stored by the retired software product.
  - Develop and document the steps required for the retirement.

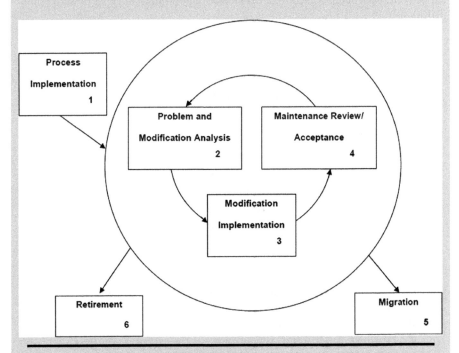

Figure 8.3   Software maintenance process.

EXERCISE 8.2: DH MAINTENANCE PROCESS

After Michel's tutorial on software maintenance processes, the team discussed what sort of maintenance process would be needed for the *DigitalHome* System. The exercise, described at the end of this chapter, deals with this issue.

EXERCISE 8.3: DH MAINTENANCE COSTS

After Michel's tutorials on software maintenance fundamentals and software maintenance processes, the team discussed what maintenance costs the *DigitalHome* System would incur in its first five years of operation. The exercise, described at the end of this chapter, deals with this issue.

Maintenance Techniques

After Michel's tutorials on maintenance fundamentals and processes, the team had lots of questions about the details and mechanics of how software maintenance is carried out. So Michel agreed to deliver a tutorial on different types of maintenance techniques.

MINITUTORIAL 8.3: SOFTWARE MAINTENANCE TECHNIQUES

IMPACT ANALYSIS

Impact analysis is a technique used to analyze the impact of change to an existing software system. Those maintaining the software must have knowledge of its requirements, design, and construction. This allows them to determine which other systems will be affected by a software change request and to estimate the resources needed to make the change. Impact analysis also includes analysis of the risk of making the change.

IEEE 14764 lists the following impact analysis tasks:

- analyze the modification requests (MRs) and/or problem reports (PRs);
- replicate or verify the problem;
- develop options for implementing the modification;
- document the MR/PR, the results, and the execution options;
- and obtain approval for the selected modification option.

RE-ENGINEERING

A *legacy system* is a system that is crucial to the enterprise, which often has the following properties: it is very old, extensively modified over the years, based on old technology, members of original development team are no longer available, and it is very expensive to maintain. Solutions include:

- Live with it
- Re-implement from scratch
- Re-engineer current system

Software re-engineering involves the analysis and re-building of an existing software system into a new reconstituted form. The following are some of the problems that may occur in a re-engineering effort:

- takes a long time;
- has a high cost;
- diverts resources that could be used in other critical areas;
- and requires a change in corporate culture.

For example, the re-engineering of the Traffic Management System to accommodate changing traffic patterns and the incorporation of new traffic management devices might be overly costly and divert limited city funds away from road maintenance.

Figure 8.4 shows the elements of a re-engineering process. Each element of the process is discussed in the following sections.

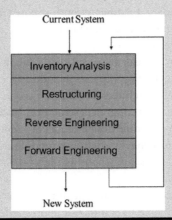

Figure 8.4 Re-engineering process model.

Inventory Analysis

Inventory analysis involves the following activities:

- Collect and organize the following information about active applications:
  - Name
  - Year created number of changes
  - Effort to make changes
  - Number of users
  - Operational costs
  - Business criticality
  - etc.
- Evaluate and assess candidates for re engineering.
- Determine the type and magnitude of re-engineering required.

Restructuring

There are various types of "restructuring":

- *Document Restructuring –*
  - Weak documentation is a trademark of legacy systems.
  - Evaluate current state of documentation using static analysis tools to produce graphical, textual, and tabular information about the source code.
  - Three options are used:
    - No changes
    - Create full documentation
    - Only update documentation for most critical parts (parts with largest maintenance potential)
- *Design Restructuring –*
  - Software with a weak architecture is difficult to adapt and enhance.
  - Design restructuring activities include:
  - analyze current architecture features: scope and locality, visibility and information hiding, data structures, and security issues;
  - and redesign a module design model by identifying and defining its attributes and behavior, defining module relationships, determining module access, and choosing physical data structures.
- *Code Restructuring –*
  - This is the most common type of re-engineering. Some legacy systems have good architecture, but poorly coded/designed modules.
  - Code restructuring activities include:

- analyze code with a restructuring tool;
- note violations of "structured" code and correct violations;
- review and test code (including regression tests);
- and update internal documentation.

Reverse Engineering

Reverse engineering is the process of analyzing a program to create a representation of the program at a higher level of abstraction than the source code. For example, create an architectural diagram from source code.

The goal of reverse engineering is to use tools and techniques to automatically extract data, architectural, and procedural design information from existing software. Current successful reverse engineering activities rely on interaction between humans and tools.

Forward Engineering

First reverse engineering is used to recover design information from a software product.

Then the software engineering process is used to create a new version of the product that is of higher quality and is easier to maintain. The new version may include new requirements and use new technology.

EXERCISE 8.4: DH RE-ENGINEERING

After Michel's tutorial on software maintenance techniques, the team discussed the scope and magnitude of future re-engineering for the *DigitalHome* System, and how the technique would be carried out. The exercise, described at the end of this chapter, deals with this issue.

Case Study Exercises

EXERCISE 8.1: IDENTIFYING POTENTIAL *DIGITALHOME* MAINTENANCE PROBLEMS

Scenario

After Michel's tutorial on software maintenance fundamentals, the team reviewed IEEE Std 14764-2006, and the DH design diagram in Figure 8.5. Then the team began a discussion of potential maintenance problems.

LEARNING OBJECTIVES

Upon completion of this module, students will have increased ability to:

- Describe the types of maintenance problems.
- Use design documents to identify potential maintenance problems.
- Understand the scope and importance of software maintenance.

EXERCISE DESCRIPTION

1. This exercise follows a lecture on software maintenance fundamentals.
2. As preparation for this exercise students are required to read the following:
   - DH Beginning Scenario
   - *DigitalHome* Need Statement
   - *DigitalHome* Software Requirements Specification
   - IEEE Std 14764-2006
3. Students are grouped into teams of 3–4.
4. Each team reviews and discusses the DH Design Diagram in Figure 8.5.
5. Each team discusses the different types of maintenance problems and identifies two possibilities in each category (corrective, adaptive, perfective, and preventive).
6. The team presents its findings to the rest of the class.

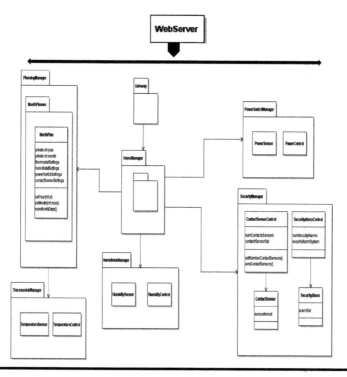

Figure 8.5   DH design diagram.

EXERCISE 8.2: DETERMINING A *DIGITALHOME* MAINTENANCE PROCESS

SCENARIO

After Michel's tutorial on software maintenance processes, the team discussed what sort of maintenance process would be needed for the *DigitalHome* System.

LEARNING OBJECTIVES

Upon completion of this module, students will have increased ability to:

- Describe and discuss the elements of a software maintenance process.
- Create a software maintenance process.
- Understand the scope and importance of software maintenance.

EXERCISE DESCRIPTION

1. This exercise follows a lecture on the software maintenance processes.
2. As preparation for this exercise students are required to read the following:
   - DH Beginning Scenario
   - *DigitalHome* Need Statement
   - *DigitalHome* Software Requirements Specification
   - IEEE Std 14764-2006, Software Engineering-System Life Cycle Processes-Maintenance [IEEE 14764].
3. Students are grouped into teams of 3–4.
4. Each team reviews and discusses [IEEE 14764].
5. Each team decides what sort of maintenance process would be appropriate for *DigitalHome.* They consider some of the following questions:
   - What size maintenance team do you think will be needed for *DigitalHome*?
   - How should "Problem and Modification Analysis" be handled? For example, how would a home owner generate a MR or PR?
   - How would a modification be reviewed?
   - How long to "Retirement" for *DigitalHome*?
6. The team presents its findings to the rest of the class.

EXERCISE 8.3: ESTIMATING *DIGITALHOME* MAINTENANCE COSTS

SCENARIO

After the team reviewed IEEE Std 14764-2006, they discussed the costs that would be incurred in maintaining the *DigitalHome* System in its first five years.

LEARNING OBJECTIVES

Upon completion of this module, students will have increased ability to:

- Describe the importance of cost in the software maintenance process.
- Estimate the cost for software maintenance.
- Understand the scope and importance of software maintenance.

EXERCISE DESCRIPTION

1. This exercise follows a lecture on software maintenance fundamentals.
2. As preparation for this exercise students are required to read the following:
   - DH Beginning Scenario
   - *DigitalHome* Need Statement
   - *DigitalHome* Software Requirements Specification
   - IEEE Std 14764-2006
3. Students are grouped into teams of 3–4.
4. Each team reviews and discusses the operation of the system in its first five years. They will make estimates of the number and types of maintenance problems, which will be encountered in the first five years of operation. For example, how many faults will be experienced and what types of enhancements might be desired.
5. Each team will estimate the average cost of finding and removing a defect, and the average cost of implementing a system enhancement. Then the team will compute the total estimated cost of maintenance over five years.
6. The team presents its findings to the rest of the class.

EXERCISE 8.4: ENVISIONING AN *DIGITALHOME* RE-ENGINEERING EFFORT

SCENARIO

After Michel's tutorial on software maintenance techniques, the team discussed the scope and magnitude of future re-engineering for the *DigitalHome* System, and how the technique would be carried out.

LEARNING OBJECTIVES

Upon completion of this module, students will have increased ability to:

- Describe and discuss the elements of a re-engineering effort for a software system.
- Determine the resources and plans needed for a re-engineering effort for a software system.
- Understand the scope and importance of software maintenance.

EXERCISE DESCRIPTION

1. This exercise follows a lecture on the *DigitalHome* Software Maintenance.
2. As preparation for this exercise students are required to read the following:
   - DH Beginning Scenario
   - *DigitalHome* Need Statement
   - *DigitalHome* Software Requirements Specification
   - IEEE Std 14764-2006, Software Engineering-System Life Cycle Processes-Maintenance [IEEE 14764].

3. Students are grouped into teams of 3–4.
4. Each team reviews and discusses [IEEE 14764].
5. Each team discusses how *DigitalHome* might evolve over the next decade and when and why a re-engineering effort would be necessary. They consider some of the following questions:
   - How would you determine when a
   - *DigitalHome* re-engineering effort would be needed?
   - What do you envision the scope of the effort to be?
   - What resources would be needed?
   - How long do you think the effort would take?
6. The team presents its findings to the rest of the class.

Chapter 9

Acting Ethically and Professionally

Software Engineering Professional Issues

In August 202X, *HomeOwner* held its annual strategic planning retreat. One of the points made by CEO Robert "Red" Sharpson was that he wanted all employees to act in an honorable and professional manner. He challenged all department heads to develop and institute workshops for this purpose. In September 202X, during the *DigitalHome* project launch, Jose Ortiz delivered a tutorial to address the CEO's challenge.

MINITUTORIAL 9.1: SOFTWARE DEVELOPMENT PROFESSIONAL ISSUES

Jose began his tutorial with a discussion of the software engineering profession.

THE SOFTWARE ENGINEERING PROFESSION

In some areas there have been significant advancement in software engineering practice; however, serious problems in software development associated with cost, quality, and schedule still remain.

There has been much improvement in the professional preparation for software engineering; but, many academic programs in computing devote insufficient attention to software engineering issues. Although improving, many organizations still do not use "best practices" (or worst, they are not

---

aware of best practices). The software engineering profession is still viewed as immature. Some even dispute that it is a profession.

James Brooks in [Brooks 1995] cautioned us about the profession:

> The tar pit of software engineering will continue to be sticky for a long time to come ... software systems are perhaps the most intricate of man's handiworks. This complex craft will demand our continual development of the discipline, our learning to compose in larger units, our best use of new tools, our best adaptation of proven engineering management methods, liberal application of common sense, and a God-given humility to recognize our fallibility and limitations.

In [Ford 1996], the authors characterize and model the evolution and maturation of the software engineering profession. A *profession* is characterized as a calling that requires specialized preparation, validation by recognized bodies, continued professional development, observance of a code of ethics, and involvement with a professional society. Figure 9.1 depicts these elements of a profession. The legal and medical professions are classic examples

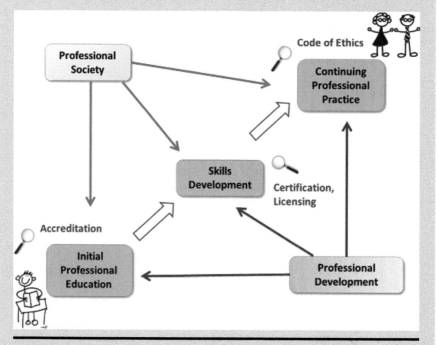

Figure 9.1 Components of a profession.

of this characterization of a profession. Software engineering is much younger profession and not as fully mature as more traditional professions (such as law, medicine, and other fields of engineering).

ACCREDITATION

Accreditation is a designation that an organization or business has met a combination of standards and abilities that are put in place for public safety, welfare, and confidence.

Engineering education accreditation standards and criteria, and curriculum guidance are provided by a number of accreditation organizations across a variety of nations and regions:

- Accreditation Board for Engineering and Technology (ABET (http://www.abet.org/)
- British Computer Society (BCS) (http://www1.bcs.org.uk/)
- Canadian Engineering Accreditation Board (CEAB)(http://www.ccpe.ca/)
- Japan Accreditation Board for Engineering (JABEE) (http://www.jabee.org/)
- European Network for Accreditation of Engineering Education (http://www.enaee.eu/)

In the United States, ABET is the primary engineering and computing accreditation organization. The purpose of the ABET (https://www.abet.org) is as follows:

- Organize and carry out a comprehensive process of accreditation of pertinent programs leading to degrees and to assist academic institutions in planning their educational programs.
- Promote the intellectual development of those interested in engineering, technology, computing, and applied science professions, and provide technical assistance to agencies having professional regulatory authority applicable to accreditation.

Engineering and computing programs are accredited based on a set of criteria for students, curriculum, faculty, facilities, institution support, and the assessment and evaluation of these criteria. The Engineering Accreditation Commission (EAC) of ABET specifies that engineering programs (e.g., Software Engineering) must demonstrate their ability to prepare graduates to achieve the following outcomes:

1. an ability to identify, formulate, and solve complex engineering problems by applying principles of engineering, science, and mathematics
2. an ability to apply engineering design to produce solutions that meet specified needs with consideration of public health, safety, and welfare, as well as global, cultural, social, environmental, and economic factors
3. an ability to communicate effectively with a range of audiences
4. an ability to recognize ethical and professional responsibilities in engineering situations and make informed judgments, which must consider the impact of engineering solutions in global, economic, environmental, and societal contexts
5. an ability to function effectively on a team whose members together provide leadership, create a collaborative and inclusive environment, establish goals, plan tasks, and meet objectives
6. an ability to develop and conduct appropriate experimentation, analyze and interpret data, and use engineering judgment to draw conclusions
7. an ability to acquire and apply new knowledge as needed using appropriate learning strategies.

In addition, software engineering programs have the following additional criteria:

■ Curriculum

The curriculum must provide both breadth and depth across the range of engineering and computer science topics implied by the title and objectives of the program.

The curriculum must include computing fundamentals, software design and construction, requirements analysis, security, verification, and validation; software engineering processes and tools appropriate for the development of complex software systems; and discrete mathematics, probability, and statistics, with applications appropriate to software engineering.

■ Faculty

A software engineering program must demonstrate that faculty members teaching core software engineering topics have an understanding of professional practice in software engineering and maintain currency in their areas of professional or scholarly specialization.

LICENSING AND CERTIFICATION

Licensing is a mandatory process administered by a governmental authority. In the United States, doctors, lawyers, engineers, and barbers are licensed by individual states. Licensing typically requires passing a licensing exam and some minimum experiences as a practicing engineer. Currently few states license software engineers, the first was the state of Texas.

Certification is a voluntary process administered by a profession that certifies that an individual has a certain level of competency. For software engineers, the IEEE Computer Society administers a Certified Software Development Professional (CSDP) (http://www.computer.org/certification/) program. Candidates seeking the CSDP certification should have completed a minimum of two years of college education in computer science or the equivalent in a related field and two years of relevant experience in industry. Certification requires successful completion of Certificates of Proficiency in the following four key knowledge areas: Software Requirements, Software Design, Software Construction, Software Testing, and the successful completion of two applied modules.

PROFESSIONAL SOCIETIES

A professional society is a nonprofit organization seeking to further a particular profession, the interests of individuals engaged in that profession, and the public interest. The following are examples of prominent professional societies:

- *ABA (American Bar Association) – lawyers*
- *AMA (American Medical Association) – physicians*
- *ASME (American Society of Mechanical engineers) – mechanical engineers*
- *IEEE (Institute of Electrical and Electronic Engineers) – electrical engineers*

There are no professional societies solely devoted to software engineering. Two societies are associated with software engineering interests and practices:

- Association for Computing Machinery (ACM) http://acm.org/
- IEEE Computer Society (IEEE-CS) http://www.computer.org/

The ACM has a Special Interest Group on Software Engineering (SIGSOFT), which "seeks to improve our ability to engineer software by stimulating interaction among practitioners, researchers, and educators; by fostering the professional development of software engineers; and by representing software engineers to professional, legal, and political entities". The scope of the Group's specialty is software engineering methods which are related to the design and construction of high-quality software systems. Topics of interest include programming techniques, methodologies for system design and implementation, debugging and testing, validation and verification, program portability, management of software development, and specification techniques.

The Constitution of the IEEE Computer Society states:

> The purposes of the Society shall be scientific, literary, and educational in character. The Society shall strive to advance the theory, practice, and application of computer and information processing science and technology and shall maintain a high professional standing among its members. The Society shall promote cooperation and exchange of technical information among its members and to this end shall hold meetings for the presentation and discussion of technical papers, shall support lifelong professional education and certification, shall develop standards, shall publish technical journals, shall provide technical and professional products and services, and shall through its organization and other appropriate means provide for the needs of its members.

EXERCISE 9.1: HIRING A NEW DH TEAM MEMBER

After about two months of work on the DH project, Jose Ortiz and Disha Chandra discussed the progress of the team and decided they could use one more team member.

In this exercise, student groups, simulating the DH Team, discuss what sort of professional capabilities the new hire should have and how they will assess applicants for the new position.

EXERCISE 9.2: ABET STUDENT OUTCOMES

In his tutorial on software engineering professionalism, Jose Ortiz asked the DH Team to rate their preparation using the ABET Student Outcomes (1)–(7). The exercise, described at the end of this chapter, deals with this issue.

Code of Ethics and Professional Conduct

After Jose Ortiz had completed his tutorial on Software Development Professional Issues, Disha Chandra pointed out that in his discussion of Figure 9.1 on the Components of a Profession he had mentioned the importance of a Code of Ethics; but he had not discussed the topic. Jose said, "good point", and that he was preparing a follow-up tutorial on this subject.

MINITUTORIAL 9.2: CODE OF ETHICS

Jose began his discussion with a quote from [Ford 1996]: "In order to ensure that its practitioners behave in a responsible manner, many professions have adopted a code of ethics (sometimes called a code of conduct or code of practice)". Some of the reasons for adopting such codes are as follows:

- Professionals are concerned with the effect of their practice on society.
- Professionals typically set higher standards for themselves than society requires.
- For the privilege of being viewed as a "professional", members of a profession accept responsibility for their actions.
- Professionals want to be viewed as acting in an ethical, responsible manner.

As was mentioned earlier, ABET has a student outcome, which states that graduates must have "an ability to recognize ethical and professional responsibilities in engineering situations and make informed judgments, which must consider the impact of engineering solutions in global, economic, environmental, and societal contexts". A code of ethics supports achievement of such an outcome.

Most established professions have a code of ethics. Because of their maturity and their impact on the public, medicine and law, have well-established codes of ethics. The Hippocratic Oath (400 B.C.) [Ford 1996] provides an early ethical standard for physicians:

> I will follow that system of regimen which, according to my ability and judgment, I consider for the benefit of my patients, and abstain from whatever is deleterious and mischievous ...

While I continue to keep this Oath unviolated, may it be granted to me to enjoy life and the practice of the art, respected by all men, in all times! But should I trespass and violate this Oath, may the reverse be my lot!

The IEEE Code Ethics, which covers electrical and computing disciplines, states (https://www.ieee.org):

We, the members of the IEEE, in recognition of the importance of our technologies in affecting the quality of life throughout the world, and in accepting a personal obligation to our profession, its members, and the communities we serve, do hereby commit ourselves to the highest ethical and professional conduct ...:

In 1999, the *Software Engineering Code of Ethics and Professional Practice* (SE Code) was developed by the ACM/IEEE-CS Joint Task Force on Software Engineering Ethics and Professional Practices, and jointly approved by the ACM and the IEEE-CS as the standard for teaching and practicing software engineering [ACM 1999].

(https://ethics.acm.org/code-of-ethics/software-engineering-code/)

The SE Code (short version) states:

Software engineers shall commit themselves to making the analysis, specification, design, development, testing and maintenance of software a beneficial and respected profession. In accordance with their commitment to the health, safety and welfare of the public, software engineers shall adhere to the following Eight Principles:

- *Public* – Software engineers shall act consistently with the public interest.
- *Client and Employer* – Software engineers shall act in a manner that is in the best interests of their client and employer consistent with the public interest.
- *Product* – Software engineers shall ensure that their products and related modifications meet the highest professional standards possible.
- *Judgment* – Software engineers shall maintain integrity and independence in their professional judgment.
- *Management* – Software engineering managers and leaders shall subscribe to and promote an ethical approach to the management of software development and maintenance.
- *Profession* – Software engineers shall advance the integrity and reputation of the profession consistent with the public interest.
- *Colleagues* – Software engineers shall be fair to and supportive of their colleagues.

■ *Self* — Software engineers shall participate in lifelong learning regarding the practice of their profession and shall promote an ethical approach to the practice of the profession.

Codes of ethics have been created in response to actual or anticipated ethical conflicts. Considered in a vacuum, many codes of ethics would be difficult to comprehend or interpret. It is only in the context of real life and real ethical ambiguity that the codes take on any meaning.

**EXERCISE 9.3: THE DH TEAM FACES
SOME ETHICAL DILEMMAS**

In his tutorial on codes of ethics Jose Ortiz asserted that such codes are best understood in the context of real life. The exercise, described at the end of this chapter, deals with this issue. (The exercise was inspired by the ethics cases presented in [Anderson 1993] and [Weiss 1990].)

Software Development Standards

MINITUTORIAL 9.3: SOFTWARE DEVELOPMENT STANDARDS

Jose continued his tutorial with a discussion of the value and need for software development standards. These standards specify accepted and verified approaches to software development, such as the definition of technical terms, the format or organization for a software artifact, or a description of a software development technique. The importance of standards is emphasized in the ABET criteria statement that the curriculum must include "a culminating major engineering design experience that (1) incorporates appropriate *engineering standards* and multiple constraints, and (2) is based on the knowledge and skills acquired in earlier course work".

The Software & Systems Engineering Standards Committee (S2ESC) of the IEEE Computer Society is a leader in the development of software engineering standards. These standards address topics such as requirements analysis and specification, software design, software testing, verification and validation, software maintenance, measurement, and plans, and documentation aspects of software engineering projects. Knowledgeable and experienced individuals from industry, government, and academia participate in developing these standards.

Table 9.1 provides a short list of some the most widely used IEEE-CS standards. *The Road Map to Software Engineering: A Standards-Based Guide* [Moore 2006] provides an overview of software and systems engineering standards developed by the S2ESC and the IEEE-CS.

EXERCISE 9.4: DETERMINING DH STANDARDS

After his tutorial on software engineering standards, Jose Ortiz asked the DH Team to consider which standards were appropriate for the DH project. The exercise, described at the end of this chapter, deals with this issue.

Table 9.1   IEEE-CS Software Engineering Standards

ID	Title	Purpose
ISO/IEC/IEEE 24765	Systems and software engineering Vocabulary	The vocabulary defines terms in the field of systems and Software Engineering. E.g., *equivalent faults*: Two or more faults that result in the same failure mode.
IEEE Std 12207-2008	Systems and software engineering Software life cycle processes	Provides a defined set of processes to facilitate communication among acquirers, suppliers, and other stakeholders in the life cycle of a software product.
IEEE Std 830-1998	Recommended Practice for Software Requirements Specifications	Provides recommended practice for writing software requirements specifications. It describes the content and qualities of a good software requirements speciation (SRS) and presents several sample SRS outlines.
IEEE Std 1016™-2009	Systems Design – Software Design Descriptions	Specifies the information, content, and organization for software design descriptions (SDDs).

(Continued)

Table 9.1 (*Continued*)   IEEE-CS Software Engineering Standards

ID	Title	Purpose
IEEE Std 1012™-2012	System and Software Verification and Validation (V&V)	Establishes a common framework for system and software V&V processes, and activities, and defines the V&V tasks, required inputs, and required outputs in each life cycle process.
IEEE Std 14764-2006	Software Engineering – Software Life Cycle Processes – Maintenance	Describes an iterative process for managing and executing software maintenance activities.

Software Legal Issues

MINITUTORIAL 9.4: SOFTWARE LEGAL ISSUES

Jose completed his tutorial with a discussion of software legal issues. The development and use of software can lead to a number of legal problems: violation of the rights of the developer organization and/or their developers; negligence that causes harm to customers or users; and contract violations by a development organization, a vendor, or a customer. Jose covered the following legal issues (influenced by material from [Armour 1993]) that are relevant to software development:

INTELLECTUAL PROPERTY

Intellectual property (IP) is a category of property that includes intangible creations of the human intellect, and primarily encompasses copyrights, patents, and trade secrets. For software, IP includes all work that is created in the process of developing a software product – for example, the requirements specification, design documents, source code, graphical user interfaces, test plans, logos, and names.

COPYRIGHTS

Developers and software companies use a *software copyright* to prevent the unauthorized copying of their software. Free and open source licenses also rely on copyright law to enforce their terms. In the United States, computer programs are copyrighted as literary works of creation.

PATENTS

A *patent* is a set of exclusionary rights granted by a government to a patent holder for a limited period, usually 20 years. These rights are granted to patent applicants in exchange for their disclosure of the inventions. Once a patent is granted in a given country, no person may make, use, sell or import/export the claimed invention in that country without the permission of the patent holder.

A *software patent* is a patent on a piece of software, such as a computer program, libraries, user interface, or algorithm. The rules for awarding software patents vary by country; in the United States, patent law excludes "abstract ideas", and this has been used to refuse some patents involving software.

TRADE SECRETS

A *trade secret* is a practice, process, design, instrument, formula, pattern, or a set of information not generally known by others, by which a business can obtain an economic advantage over competitors or customers. Trade secret law provides protection for a software product (e.g., its design and source code). It simply requires that you take reasonable efforts to keep the item secret. There are no formalities, such as filing with a government agency, required.

The legal status of a trade secret is a protected intellectual property right if the owner can prove the trade secret was not generally known and reasonable steps were taken to preserve its secrecy.

LEGAL PROBLEMS

The abuse of intellectual property rights such as violation of a patent or a copyright, or inappropriate disclosure of a trade secret can lead to legal difficulties. However, a more serious problem is the quality of a software product. Parnas and Lawford [Parnas 2003] state "Despite more than 30 years' effort to improve software quality, companies still release programs containing numerous errors. Many major products have thousands of bugs". These software defects can pose serious consequences with safety and security, and with the financial threats to customers and users. The following are issues that can place a software development organization or an individual developer in legal jeopardy:

LIABILITY

Strict software liability is the portion of law that covers damage caused by dangerous software products. In contrast to negligence, which focuses on the processes used to produce software products, strict liability focuses on the product itself and whether it contained one or more unreasonably dangerous defects.

One strategy for developers to protect themselves from legal liability would be to improve their quality assurance efforts. A corollary would be for a developing organization to highlight the importance of the Software Engineering Code of Ethics and Professional Practice, emphasizing the following:

- *Client and Employer* – Software engineers shall act in a manner that is in the best interests of their client and employer consistent with the public interest.
- *Product* – Software engineers shall ensure that their products and related modifications meet the highest professional standards possible.

NEGLIGENCE

Negligence is defined as conduct that falls below the standard established by law to protect persons against unreasonable risk of harm. Under negligence, a supplier is not responsible for every software defect that causes customer or third-party loss. Responsibility is limited to those harmful defects that it could have detected and corrected through "reasonable" quality control practices. It is the supplier's failure to practice reasonable quality assurance that constitutes negligence and that produces the liability exposure. As with the techniques limiting liability actions, reducing the possibility of negligence lawsuits depends on effective software assurance processes and software engineers adhering to the SE Code.

EXERCISE 9.5: DH SOFTWARE LEGAL ISSUES

After his tutorial on software legal issues, Jose Ortiz asked the DH Team to consider what legal issues might apply to the DH project. The exercise, described at the end of this chapter, deals with this issue.

Case Study Exercises

EXERCISE 9.1: HIRING A NEW *DIGITALHOME* TEAM MEMBER

Scenario

After two months of work on the DH project, Jose Ortiz and Disha Chandra discussed the progress of the team and decided they could use one more team member. The DH Team discussed what sort of professional capabilities the new hire should have and how they would assess applicants for the new position.

Learning Objectives

Upon completion of this exercise, students will have increased ability to:

- Describe the various attributes of a professional.
- Use knowledge of professionalism to select a team member.
- Assess their own professional strengths and weaknesses.

Exercise Description

1. This exercise follows a lecture on software engineering professionalism.
2. As preparation for this exercise students are required to read the following:
   - DH Beginning Scenario
   - *DigitalHome* Need Statement
   - Table 2.1: *DigitalHome* Development Process Script
3. Students are grouped into teams of 3–4.
4. Each team reviews and discusses the personnel needs of the DH Project, and what professional capabilities a new member should possess.
5. Each team discusses how an applicant for a new position should be assessed.
6. The team presents its findings to the rest of the class.

EXERCISE 9.2: ABET STUDENT OUTCOMES

Scenario

In his tutorial on software engineering professionalism, Jose Ortiz asked each member of the DH Team to rate their college preparation using the ABET Student Outcomes (1)–(7), In this exercise, each student carries out such a rating.

LEARNING OBJECTIVES

Upon completion of this module students will have increased ability to:

- Describe what preparation one needs to become a software engineering professional.
- Assess how well they are being prepared to become a software engineering professional.
- Describe ABET student outcomes.

EXERCISE DESCRIPTION

1. This exercise follows a lecture on software engineering professionalism.
2. As preparation for this exercise, students review the EAC ABET accreditation material at http://www.abet.org/accreditation/accreditation-criteria/ for the current academic year.
3. Students are grouped into teams of 3–4.
4. Each team reviews and discusses the ABET EAC student outcome statements.
5. Each member of the team rates the degree to which they think each outcome is achieved in their program on a scale of 1–5. (1 = poorly achieved, 5 = highly achieved).
6. The team presents its ratings to the class.

EXERCISE 9.3: THE DH TEAM FACES SOME ETHICAL DILEMMAS

SCENARIO

In his tutorial on software engineering professionalism, Jose Ortiz asserted that codes of ethics should be viewed in terms of real-world problems. In this exercise, the DH Team assesses several software engineering ethical cases.

LEARNING OBJECTIVES

Upon completion of this module students will have increased ability to:

- Describe the elements of the Software Engineering Code of Ethics and Professional Practice (SE Code).
- Assess ethical dilemmas encountered in the software engineering profession.
- Make ethical decisions using the SE Code.

EXERCISE DESCRIPTION

1. This exercise follows a lecture on software engineering professionalism.
2. Students are grouped into teams of 3–4 people.
3. Each team reviews and discusses the SE Code (https://ethics.acm.org/code-of-ethics/software-engineering-code/).
4. Each team is assigned one of the below ethical cases. Then the team evaluates the ethical and professional issues involved in the case as follows:
   - Cites which parts of the SE Code are applicable.
   - Characterizes the professional ethics of each character involved in the case.

- Assesses the ethics of the case using the following criteria:

    Unethical does not conform to an appropriate standard of ethical/professional conduct.

    Not unethical does not violate an appropriate standard of ethical/professional conduct.

    No ethics issue does not involve an appropriate standard of ethical/professional conduct.

5. The team presents its findings to the class.

Case 1

When Georgia Magee was employed at Volcanic Power, she worked on the software for the development of a power monitoring product for new homes built by the Homestead Construction Company. She visited several construction sites to determine how best to connect the power monitoring unit to the house's electrical system. Georgia noticed that some of the electrical wiring was not properly installed and might be unsafe. Georgia reported this problem to Ed, her supervisor at Volcanic Power. Ed tells her "this is Homestead's problem and we need to stay out of it". Georgia says "okay".

Case 2

Early in his career Michel Jackson worked for *AirLoft*, an aerospace company. *AirLoft* was developing a wing section to be used in the development of new Air Force transport airplane. Michel's job was to collect and analyze data taken from the wind tunnel tests. He developed software to assist in the collection and analysis of the test data. Christine, the Project Manager for the Air Force contract, had scheduled a major design review of the wing section project in the coming week, and Michel was scheduled to present his analysis of the wind tunnel tests. However, when Michel showed the analysis results to Christine, she asked him to perform some edits to present the data in a more favorable light. Michel objected, but eventually relented after a testy exchange with Christine.

Case 3

When Disha Chandra worked at *SoftMedic*, she developed a hospital management system for a hospital customer. She included several innovative features and was highly praised. In the following year, she was an observer at a design review of a new *SoftMedic* health tracking product, being developed by Eric, a colleague at *SoftMedic*. In the design review, she noticed that several of her original design ideas for the hospital management system had been incorporated in the new health tracking product. She asked Eric about this and he admitted that he had "borrowed" some of her ideas. She was upset that Eric had not asked her about this and had not given her any credit.

Case 4

When Massood Zewail was an intern at *MacroSoft Corporation*, he was invited to be part of software team developing a student information system for Skyline University. One of the first tasks for the team was to collect and specify the requirements for the system. The Team Leader, Henry, scheduled a meeting with the Director of Student Records at Skyline and several of her associates. During the meeting Henry and other team members asked questions about what capabilities the new system should have. After about an hour, Henry ended the meeting and thanked the Student Records personnel for their help. After the team returned to *MacroSoft* offices, they met and discussed the results. Henry said he was pleased thought they were ready write a requirements specification for the proposed system.

Massood was concerned – his software engineering courses had emphasized the importance of the proper analysis and specification of software requirements and how projects could go astray if the specification did not accurately and completely capture the requirements. He spoke up and stated that he thought there was more the team could do: interview users of the system (admin staff, faculty, and students); and develop a user interface prototype and observe user interaction. Henry dismissed Massood concerns with "Massood this is real world, not a classroom. What you suggest costs time and money, and our budget is limited".

EXERCISE 9.4: DETERMINING DH STANDARDS

SCENARIO

After his tutorial on software engineering standards, Jose Ortiz asked the DH Team to consider which standards were appropriate for the DH project.

LEARNING OBJECTIVES

Upon completion of this module students will have increased ability to:

- Describe the importance of software development standards.
- Explain the contents of some of the IEEE-CS software standards.
- Determine which standards are appropriate for a software project.

EXERCISE DESCRIPTION

1. This exercise follows a lecture on software engineering professionalism.
2. Students are grouped into teams of 3–4.
3. Each team reviews and discusses Table 9.1 and examines the list of standards at http://users.encs.concordia.ca/~eceweb/capstone/Software/std_list.htm.
4. Each team selects which standards are appropriate for the DH project using the below considerations.
   - The team is developing *DigitalHome* prototype, which will be the foundation for future development.

- In development of the prototype, *DigitalHomeOwner* has two primary client groups: (1) potential *DigitalHome* users and (2) the upper management of *HomeOwner*. These clients will be part of an acceptance testing process for the prototype.
- Jose Ortiz will deliver a report on the *DigitalHome* efforts at the August 201Y strategic planning retreat.
5. The team presents its findings to the class.

EXERCISE 9.5: DH SOFTWARE LEGAL ISSUES

SCENARIO

After his tutorial on software legal issues, Jose Ortiz asked the DH Team what legal issues might apply to the DH project.

LEARNING OBJECTIVES

Upon completion of this module students will have increased ability to:

- Explain some key legal issues related to software development.
- Describe the importance of software legal issues for developers and customers.
- Determine which legal issues should be considered for the DH Project.

EXERCISE DESCRIPTION

1. This exercise follows a lecture on software engineering professionalism.
2. Students are grouped into teams of 3–4.
3. Each team reviews and discusses the publication Software Product Liability (https://resources.sei.cmu.edu/asset_files/TechnicalReport/1993_005_001_16187.pdf).
4. Each team answers the following questions:
   - Are there any elements of the DH project, which should be considered as trade secrets, or for patent or copyright protection?
   - In order to limit future problems with liability or negligence are there changes you would recommend in Table 2.1: *DigitalHome* Development Process Script?
5. The team presents its findings to the class.

Chapter 10

Using the Scrum Development Process

Scrum Process Overview

As part of the August 202X, *HomeOwner* annual strategic planning retreat, the development team delivered a number of MiniTutorials explaining the plan-driven process, however, Massood had doubts about whether this was a good approach or not. He raised his concern to Jose and suggested perhaps an agile process would be a better approach for the development of the prototype. Given the fact that no one on the team had an in-depth knowledge of agile processes, Jose asked Massood to develop a series of short MiniTutorials about agile processes and deliver them to the team. Massood accepted the challenge and decided to prepare a number of MiniTutorials about the Scrum process.

NOTE: *There are a lot of commonalities associated with the plan-driven and agile process activities; therefore, what has been covered in the previous nine chapters are still applicable to the agile processes. For example, the discussion of forming a team in Chapter 2 would still be valuable when you are forming an agile team, or the discussion of software design, architecture, and design pattern is still applicable to the agile process (with some minor modifications).*

MINITUTORIAL 10.1:   SCRUM PROCESS OVERVIEW

Massood began his tutorial with an overview of the Scrum process.

The Scrum Guide [Schwaber 2017] is a framework to facilitate productivity by prioritizing tasks with highest value and by working in short time

---

increments within a "inspect and adapt" framework. The Scrum framework is founded on empirical process control theory which is based on three principles:

- *Transparency* – Significant aspects of the process must be visible to those responsible for the outcome.
- *Inspection* – The artifacts and the processes must be frequently inspected to verify the progress toward the goals that are made, and any undesirable variance from the goals and the process are identified.
- *Adaptation* – If any variation in product and/or process is outside the accepted range, then appropriate adjustment must be taken as soon as possible in order to prevent further deviation.

The Scrum framework uses a cross-functional team to develop a product (complete a project) in an iterative incremental manner in order to optimize predictability and control risks. This means tasks are completed over series of iterations, where at the completion of each iteration, an increment of the product is delivered. Each iteration is referred to as a Sprint and its duration can be as small as couple of days, or as long as multiple weeks, but the most common iteration size is two weeks. One of the fundamental principles in the Scrum framework is that the duration of the sprint is "time boxed", which means the sprint will end on the designated date (end of the sprint), whether the tasks assigned to that sprint are complete or not. In addition to the sprints, there are number of artifacts, meetings (events), and roles that are the cornerstone of the Scrum process. There are four types of meetings (events) in the scrum framework: Sprint Planning, Daily Scrum, Sprint Review, and Sprint Retrospective (described later). Each has specific goals, objectives, and a set of deliverables, and they all are time boxed, in order to control the meeting and prevent waste of time. There are a number of artifacts in the Scrum framework. Some of the most common ones include Product Backlog (backlog items is a description of functionality, which is covered in MiniTutorial 10.2), Release Backlog, Sprint Backlog, and Sprint Burndown Chart (described in MiniTutorial 10.3). Finally, there are three roles in the Scrum framework:

- *Product Owner* – The product owner role is in some way similar to the product manager in the traditional development process, with one major difference. In the Scrum framework, the product owner is actively involved with the team by continuous prioritization of the backlog items and reviewing the deliverable product at the completion of each sprint. The product owner is responsible for maximizing the Return on Investment (ROI) resulting from the work that is performed by the team. In addition, the product owner has the responsibility to

- Serve as a liaison between the team and the customer (sometimes the product owner and the customer are the same person)
- Establishing and communicating the product vision
- Identifying the release backlog and deciding on whether to release or not
- Be the final authority for the product (i.e. does it satisfy the requirements)

■ *Scrum Master* – The scum master is the custodian and advocate for the scrum process. This means not only that (s)he is responsible for the smooth operation of the team following the scrum process. But (s)he is also responsible to interact with entities outside of the team who may or may not be familiar with the scrum process, and make sure these entities understand their interaction with the scrum team and whether these interactions are helpful or not for the success of the team, and how they can help the team by changing the interactions that are not helpful. In addition, the scrum master has the responsibility to:

- Create a safe environment for the team to share their views, and self-organization.
- Address and resolve the team's impediments (challenges) as soon as possible.
- Shield the team from external interferences and distractions.
- Take a leadership role, but not necessarily a management authority for the team.

■ *The Team* – The development team is a cross-functional team that consists of professionals who are responsible for the work that will be delivered at the end of each sprint. The development team has the authority to organize and manage their own work and no external entity may assign a specific task to a member of the team. The team size ranges from 3 to 8 people with the ideal number around 5 members per team. In addition, the team has the following characteristics:

- The ideal team member is a "T-shaped person", which means the member has an area of specialization, but they have the capability to work in other areas as needs arise.
- Accountability does not belong to one person, but to the whole team.
- There is no sub-team (i.e., specialization in different areas) in a scrum team.

In case the project requires a larger team, then it is strongly suggested to follow the Scrum Scale, which provides a scale for assessing roles, events, and enterprise artifacts, as well as the rules that bind them together [Sutherland 2019].

In addition to the above, depending on the project type, there may be other participants in the project, which include the customer, business developers, etc.

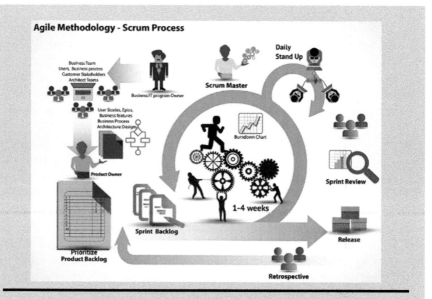

Figure 10.1 Scrum framework (©Shutterstock.com. Used with Permission.)

Figure 10.1 displays an elaborated Scrum development life cycle, in which we can identify many different activities that are conducted during the project. This has resulted in many different opinions about the different phases of the Scrum framework. These differing views have ranged from a minimum of three to as many as eight or more different distinct phases. For the purpose of this tutorial, we are recognizing three distinct categories of activities (phases if you wish). These are Initiation, Implementation, and Reflection.

INITIATION

The first set of activities in the scrum framework is the Initiation activities. The initiation activities concentrate on the development of a vision for the project, identifying and gathering appropriate project participants, identifying user stories, etc. The following are the list of activities that take place during the project initiation:

- Establish a project vision
- Identify the scope of the project
- Identify the project participants
    - Customer and corresponding stakeholders (i.e., product users, etc.)
    - Product owner (if it is different from the Customer)
    - Scrum Master
    - Cross-functional team members

- Identify initial set of the product's backlog items (themes, epics, and stories)
- Establish initial backlog prioritization
- Establish initial Release planning

IMPLEMENTATION

Once the project initiation is completed, the actual project development starts. The development process takes place over a number of sprints, where the duration of sprints are agreed upon by the team. As previously mentioned, duration of a sprint can be as small as couple of days, and as long as multiple weeks (but usually not longer than a month), with the most common duration being two weeks. During each sprint, the team will perform tasks in support of producing a releasable product increment. At the beginning of each sprint, the product owner, scrum master, and team agree on the sprint goal, and identify the backlog items that will support the accomplishment of that goal. The selection of the backlog items for the sprint is based on number of factors. In addition to the successful accomplishment of the sprint goal, additional factors that influence the selection of backlog items for the sprint include

- backlog items that have the highest priority (based on the product owner prioritization)
- Backlog items that have been estimated (estimation is covered in the next tutorial)
- backlog items that are ready to be worked on (i.e., does not depend on the completion of other backlog items that would be worked on during the later sprints)
- backlog items that are clear enough for the developers to work on
- backlog items that are possible to be delivered in one sprint, if not the backlog item should be broken into several other backlog items

Once the backlog items for the sprint (sprint backlog items) are agreed on the team will identify the tasks that need to be performed by the team in order to deliver the sprint backlog items.

REFLECTION

At the completion of each sprint, release, and the final project the team conducts a number of activities that are referred to as reflection. There are two distinct activities that are performed by the team in order to reflect on the work that has been completed:

- *Product Review* – This is where the quality of the product that is delivered as a result of a sprint, release, or the final product is being reviewed by the customer and/or other relevant stakeholders. If any variation is identified that is outside of the acceptable range, then corrective actions are identified and assigned to the product backlog.

■ *Process Review* – This is where the process that is used in order to deliver the product at the end of each sprint, release, or the final product is reviewed and opportunities for improvement are identified and are incorporated to the process for future work.

The detail activities that are conducted in a sprint are covered in the next MiniTutorials.

EXERCISE 10.1: FORMING A SCRUM TEAM

As part of the DH project in the previous chapters, a team has been formed to work on the development of the project following a plan-driven process.

In this exercise, student groups will evaluate the existing team formation against the Scrum team characteristics. The exercise should include the following evaluation:

■ Do we have the right team members to form a Scrum team?
■ If so, which team members have the qualifications to serve as Product Owner, and Scrum Master?
■ Do the remaining original team members have the appropriate qualification to be part of the team?
■ Is there anyone else that should be included as part of the Scrum organization?

Backlog Generation and Grooming

After Massood provided a brief overview of the Scrum framework, the team agreed that using Scrum process could bring in advantages over the planned-driven process. However, they all agreed that there is a need for more detailed understanding of the Scrum framework before they make the final decision on which process to follow.

MINITUTORIAL 10.2: BACKLOG GENERATION AND GROOMING

As previously mentioned, the Scrum process can be divided to three distinct phases: Initiation, Implementation, and Reflection. The initiation phase has three major goals: (1) establishing the project vision and scope, (2) forming an effective team to deliver the project, and (3) generating an initial set of Backlog Items (requirements) and their corresponding estimates to complete the Backlog items. The purpose of this tutorial is to provide an overview of the backlog generation and grooming process for a project.

BACKLOG ITEMS

There are at least four different terms that are used in Scrum framework that has reference to the requirements: (1) Backlog Items, (2) Themes, (3) Epics, or (4) Stories. Although these are all referring to the requirements, they are not all the same, which could cause a lot of confusion. To make this a bit less confusing, we use the following definitions for these terms.

Product Backlog Item (PBI) is the only official reference to the product requirements in the Scrum framework. The backlog items represent the customer's view of the product functionality. The remaining terms mentioned above refer to the different levels of complexity of the PBIs.

- *Themes (aka Use Cases, User Features)* – Big items that include many Epics. Each theme may take **many months** to complete. For example, building a fully Digital Home.
- *Epics* – Large body of works that may contain many stories (aka user stories). Each epic may take **many weeks** to complete (may or may not be completed in a single sprint). Sometimes an Epic may represent a Minimum Viable Product (MVP) or a sprint backlog item. For example, controlling the environmental setting of the Digital Home.
- *Stories (aka User Stories)* – Represents a body of work that can be completed in **many days**. User stories are usually represented in the form of a sprint backlog item. Building the control center for monitoring the temperature and humidity.
- *Tasks* – Represents the decomposed components of a story. This is dealing with "HOW" a story can be completed. A story typically could be completed in couple of hours or a day or so. For example installing temperature sensors throughout the Digital Home.

As it is shown above and represented in Figure 10.2, requirements can be represented at different levels of complexities, but once the requirement is ready to be worked on, it should be clear and concise enough so once it is brought into a sprint and it becomes part of the print backlog, it can be divided to series of tasks that the team can deliver during that sprint.

BACKLOG LIST

As previously mentioned, there are different Backlog lists in the Scrum framework, these are product backlog, sprint backlog, and an optional release backlog. The product backlog represents the overall functionality of

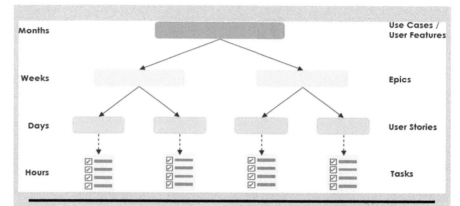

Figure 10.2   Relationship between themes, epics, stories and tasks.

the product. You can think of a product backlog list as a repository of all the things that need to be completed in the project; therefore, everyone have a stake in the generation of this backlog, and as such everyone can add backlog item to the product backlog list. It is the responsibility of the product owner to evaluate the "value" of the backlog item to the customer, and the product, and the ones that have the higherest values receive the higher priority. The product backlog items could be represented at any level of complexity as discussed previously, but before they become part of the sprint backlog, they need to be clearly defined and estimated, which takes place during the backlog grooming process (described shortly). Figure 10.3 is a graphical representation of product backlog. As it is shown, the backlog items that are going to be worked on in the next sprint are much better defined and scoped (indicated by the volume of the backlog item), and as we look into the backlog items that are potentially be worked on in later sprints, they are less defined (represented by the larger volumes).

The sprint backlog list holds all the backlogs that are going to be worked on during the current sprint. The team is responsible for the generation and grooming of the sprint backlogs. The sprint backlog items are well defined and detailed to the point that they can be decomposed to a number of tasks that team members will work on. As it is shown Figure 10.3, there is a possibility that when a product backlog item (EE) is transferred to the sprint, that is broken down to additional backlog items (E1, E2, E3, and E4) in the sprint. The team also added two additional backlog items (F and G) to the sprint backlog. These items are needed to be completed, before the sprint goal is achieved. Also, the sprint backlog item shows a block called Buffer, which will be explained in the next MiniTutorial. The third optional backlog list is called release backlog. If the product is delivered over multiple releases either due to the size or

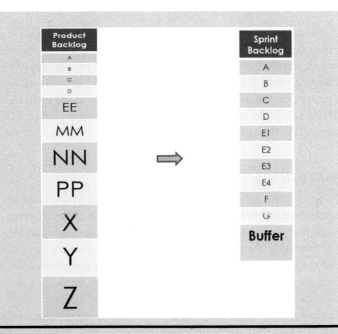

Figure 10.3   An illustration of backlog lists.

maintenance, then the team may use a release backlog list, which represents all the backlogs that will be delivered as part of the release.

An important activity in the Scrum framework is called **Backlog Refinement** or **Backlog Grooming**. The backlog refinement is the process where the product owner, and all or some members of the team review the product backlogs to make sure they are "ready/good enough" to be worked on. The characteristics of a good product backlog item are as follows:

- *Detailed* – They have the appropriate level of detail so that the developers can work on them without any misunderstanding. In another words, it is clear, concise, and unambiguous.
- *Prioritized* – One of the goals of the Scrum is to be able to deliver highest value to the customer in the most efficient way (lowest cost). In another words, make the Return on Investment (ROI) visible to the customer. One way to accomplish this is to make sure the highest priority backlog items are delivered to the customer as early as possible. Another reason for prioritization is to identify the interdependencies between the backlog items, and make sure that the backlog items that are dependent on some other ones are given lower priority than the ones that they are dependent on.

■ *Estimated* – In order to have a good plan, one should know how much effort is required to deliver a product. Therefore, before the team plans to take on any backlog item, they should have an idea about how much effort is needed to deliver that backlog item. There are a number of factors that can affect the estimate. Two such factors include, the time it takes to build/deliver that item, and the amount of technical and/or other risks that are associated with the item.

ESTIMATION

As backlog items are prepared to be worked on, they need to be evaluated based on the amount of effort they each require to be delivered. The estimation process is a critical component of the Scrum framework, as the estimates are used for the sprint planning. As previously mentioned, a successful sprint is the one in which all the sprint backlog items are delivered at the completion of the sprint. Therefore, as part of the sprint planning activity (to be discussed later), the team will decide on how much work they can deliver during the sprint. There are a number of estimation techniques that could be used (some of these were discussed in Chapter 4). The following three are the most common techniques:

■ *T-Shirt Sizing* – In this technique, each backlog item is estimated based on the size of a T-shirt (XS, S, M, L, XL). The larger the size, the more effort is required to complete that backlog item. In this technique, a team may assign a number (i.e., points, hours, days, etc.) to the T-shirt size in order to be used as part of their sprint planning. For example, one may assign the following points 2, 4, 8, 12, and 16 to each T-shirt size (XS, S, M, L, XL) respectively.
■ *Planning Poker* – In this technique, each member of the team will be given an opportunity to estimate the effort (points) required to complete a backlog item using one of the Fibonacci numbers (1, 2, 3, 5, 8, 13, 21,...). If there is an agreement on the effort, then the corresponding points are assigned as the estimate; however, if there is a wide range of opinion on the effort, then the two extreme points are ignored and another round of estimation is conducted (using the previous range as the boundary). This process will continue until either all agree on the assigned points (effort), or the average of the remaining points is assigned to the backlog item. Figure 10.4 displays and illustrates the playing poker estimation technique.
■ *Ordering Method* – In this technique, the team chooses a backlog item that is deemed to require a medium level of effort and assigns a point value (e.g., 8) to that backlog item. Then the next backlog item is compared with the backlog item that has just been assigned, in order to

Figure 10.4 An illustration of playing poker estimation. (©Agile Digest company. Used with Permission.)

estimate if it requires more effort (points) or less. This process will continue until all the backlog items are ordered from the least amount of effort to most amount of effort. Once the rank ordering is completed, then the team will evaluate the points assignment in order to make sure the assigned points are a good representation of the amount of work they require, if not, then the team may adjust points in order to make sure the backlog item with the least amount of effort having the lowest number of points, and the backlog item with the highest level of effort having the highest assigned point. Figure 10.5 illustrates the ordering method estimation technique.

Figure 10.5 An illustration of the ordering estimation.

The above-mentioned three estimation techniques are some of the more common approaches used in the Scrum framework; however, there are many other proven and experimental estimation approaches that are used by Scrum teams. The important step in estimation is to use your historical data to get better every time you go through the estimation. Having done so, you reach a point where your estimation of the effort will be as accurate as possible (i.e., within ±10%). Therefore, it is ok to miss your goals during the early sprints in the project, as long as you get better sprint after sprint. However, if you keep missing your goal the same amount or even worst during the later sprints, then you have a big problem.

EXERCISE 10.2A: THE DH TEAM BACKLOG GENERATION AND ESTIMATION

Massood thought the best way to communicate the importance of the backlog items and estimation is to come up with an exercise that everyone on the team has a good understanding of the domain, and have the team to generate the initial product backlog item and estimation. So he identified two exercises:

Exercise 10.2 – He asked the team to treat each contractor who will be involved in building Digital Home as a product backlog item. Examples of potential contractors for Digital Home would include contractors to perform the following tasks:

- Developing the communication center, through which a user can monitor and control home devices and systems.
- Setting up environmental sensors (temperature sensors, light sensors, humidity sensors, power sensors, contact sensors, water sensors, etc.), which uses wireless communication, so that sensor values can read and saved in the home database.
- Setting up a home security system, which consists of a set of contact sensors and a set of security alarms, which are activated when there is a home security breach.

The team has to generate a backlog item (list all the Digital Home contractors) and prioritize them based on which contractor have to start their work and put them in order of their starting their work.

EXERCISE 10.2B: DH BACKLOG ESTIMATION

After the completion of the tutorial on Scrum process, Massood asked the DH to use the backlog items in exercise 10.2A and conduct an estimation of the backlog items using T-shirt sizing and Playing Poker technique.

Building the Product

MINITUTORIAL 10.3: SPRINT PROCESS

The second set of activities in the scrum framework is associated with the product implementation. Sprints are where the actual implementation takes place. In the scrum framework, the product is developed over number of increments which are the outcomes of sprints. Figure 10.6 shows the overall activities associated with the sprint.

The first step in the sprint is sprint planning, where the team in collaboration with the product owner identifies the backlog items that will be worked on during the sprint. Then the team works on the delivery of those backlog items throughout the sprint. Each day the team will conduct a daily scrum, where team members inform the rest of the team about the status of what they are working on, and if there are any issues that they are facing. Throughout the sprint, the Scrum master will do what is necessary in order to make the sprint go as smooth as possible. At the end of the sprint, the increment of the product that has been completed will be presented to the customer. In the remainder of this MiniTutorial we will discuss each of these activities in more detail.

SPRINT PLANNING

Sprint planning is the first step of the sprint. It typically takes up one hour per week of sprint time. So if the sprint duration is two weeks, then the maximum amount of the time allocated to the sprint planning is two hours. There are two major activities that are associated with the sprint planning:

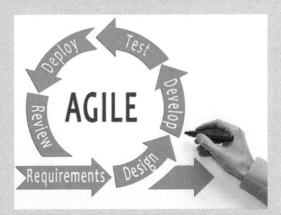

Figure 10.6   Sprint overview (©Shutterstock.com. Used with Permission.)

■ *What to Work On* – The first step is to identify what are the goals of the sprint, and what backlog items will be delivered at the completion of the sprint. In addition, the team should know how much work they can take on during the sprint.

  – The sprint backlog is the set of highest priority backlog items in the product backlog list. These backlog items have already been groomed (detailed, prioritized, and estimated) and are identified as the ones that support the overall goal of the sprint. These backlog items have been identified and agreed on with the product owner, scrum master, and the team; however, the final decision on how much work the team takes on during the sprint is decided by the team and no one else.

  – The estimated backlog items are a good start for planning which high priority backlog items have the potential to be completed by the end of sprint. However, in order to plan accurately, one should know the capability of the team in delivering the product. The capability of the team is to assess sprint performance based on the team's performance during previous sprints. **Sprint velocity** is calculated based on the amount of work that has been completed by the team over the previous sprints. Given the fact that backlog estimation is not a science, and there are always special circumstances that may affect the performance of the team during a single sprint, to have better approximation of the team capability, we will define the velocity as the average of the team velocity over three sprints. However, since the sprint velocity is calculated based on the team performance in three previous sprints, then how do we come up with the velocity for the first three sprints? In order to solve this problem, for the first three sprints, we try to rely on our historical data from previous projects. It is true that each project is different and the same is true for the performance of different teams. However, since there is no performance data for the first sprint, the best information we can use is the historical data. As we finish the first sprint, then we use the velocity data from the first sprint and the historical data to come up with the velocity for the second sprint and do the same for the third sprint.

  – Knowing how much work we can complete during a sprint (sprint velocity) is just part of the information we need for good planning. As you know life is full of surprises, and as such in order to have a more accurate plan for the sprint, we should take into consideration all these surprises. For example, if our team has some other product that was recently released, there is a possibility that some maintenance request (either to fix a bug or add new feature) will be coming our way. Therefore, there is a possibility that we may be required to spend some of our velocity during the existing sprint on taking care of those requests. In this situation, we assign a portion of our velocity into a buffer, which results into a reduction on the amount of work that can be completed in the sprint. The new effort (Velocity minus the buffer value) is referred to as the **Sprint Capacity**. The amount of work the team takes on during the sprint is based on its sprint capacity.

■ *How Do We Do It* – Once a decision is made on which backlog items will be worked on during the current sprint, the team will go through each backlog item with the goal of identifying specific tasks that are required to complete that backlog item. In this process, each task will be assigned a level of effort (points), and the hope is that the combined points for tasks associated with the backlog item will be equal to the total points assigned to that backlog item as part of the grooming process. The closer these two efforts are to each other, the more accurate our estimation process is. If there is a large difference between these two numbers, then this information should be fed back into the next sprint planning in order to improve our estimation. Once the tasks are identified, then the tasks will be posted on the **Scrum Board**. A scrum board is used by the team in order to make the progress of the team visible. Figure 10.7 shows a scrum board. There are four columns in the scrum board: (1) To do, (2) In progress, (3) Testing (optional), and (4) Done. As the column name implies, the "To do" column initially

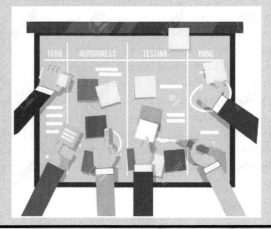

Figure 10.7   Scrum board.

shows all the tasks that need to be completed during the sprint. As the team start working on a task, the task is removed from the "To do" column, and added to the "In progress" column. For example, in the software development project, the in progress column may represent the design and development of the software product. For full transparency, the team may also include the name of the person who is working on that task. This will allow the rest of the team to know who is working on what at any moment. Once the outcome of the task is ready to be tested, the task is moved to the "testing" column. Once the task is completely tested and proven to be doing what it is supposed to do, it will be moved to the "Done" column.

DAILY SCRUM

Each morning, the team starts with a short 15 minutes stand-up meeting, called the Daily Scrum. The purpose of the daily scrum is to keep the team informed of the project progress. In addition to information sharing, the daily scrum is also used for team reorganizing (if needed). The participants in this meeting are the team members, and if needed the product owner, but no member of the management will be included in this meeting. The Scrum master is the person in charge of this meeting, and during this meeting three questions are asked. These are:

■ What did you accomplish yesterday?
■ What will you be working today?
■ Is there any impediment that you are faced with?

The answer to the first two questions will provide a snapshot of what have been completed, and what is being worked on. This information also has the potential to point to some reorganization in order to be a more efficient team. The answer to the third question reveals potential obstacles that affect the team performance and productivity. One of the responsibilities of the scrum master is to try to eliminate or at least reduce these obstacles. This may require reaching out to stakeholders who are outside of the team, and have the capability to eliminate the obstacles. For example, if the obstacle can be removed with the purchase of a tool, then the scrum master needs to reach out to the budget manager, to acquire the necessary funding to purchase the tool.

TRACKING PROGRESS

One of the three principles of the Scrum framework is transparency. There are number of ways that transparency is enforced in the process. One such approach is through the daily scrum meetings and another is through the scrum board. Another way to make the process transparent is through tracking the sprint progress. The sprint progress is represented by the **sprint burndown chart** (or burnup chart). Figure 10.8 shows a burndown chart, which is a two-dimensional graph. The X-axis represents the number of working days in the sprint, for example, if sprint has a two-week duration, then the X-axis represents ten days. The Y-axis represents the total points (efforts) that have been allocated to the sprint. As tasks are completed "done", the amount of the points associated with that task is reduced from the total points that are left to be completed. Ultimately, as the team progresses through the sprint, the number of points still to be completed is reduced, and finally it will reach zero (in a successful sprint). The comparison between the actual (red line) and projected (blue line) burndown chart provides critical information to the team on whether they are ahead or behind the "ideal" progress.

Another technique to represent the sprint progress is through the **Burnup Chart**. In this technique, the total points at the beginning of the sprint is equal to zero and as points are gained, the associated points are added to the total point gained in the sprint, therefore, the burnup chart starts at the lower left corner of the chart and raises as it goes through sprint.

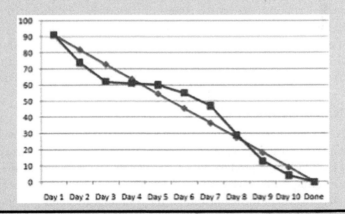

Figure 10.8   Burndown chart.

BUILDING, DOING, PRODUCING

As previously mentioned, although the scrum framework has mainly been used by software industry, it has been adopted by other industries. Therefore, no matter what industry it is used by, the ultimate goal of the sprint is to deliver an increment of the "product" that delivers some "value" to the customer. Depending on the type of "product", the team will use the accepted practices of the product development within the organization to build the increment. In the previous chapters of this book, we have covered topics such as software design, software construction, and validation and verification techniques that could potentially be used, or customized by the team to build their product increment.

EXERCISE 10.3A: DH SPRINT PLANNING

After the completion of the tutorial on Scrum process, Massood asked the DH Team to identify all the backlog items that are addressing the Digital Home temperature control.

Scrum Reflection Activities

MINITUTORIAL 10.4:   SCRUM REFLECTION ACTIVITIES

The final phase of the Scrum process is the reflection. There are two different activities that are conducted under the reflection phase, these are reviews and retrospective. Review concentrates on the review of the product, and the retrospective concentrates on the review of the process and potential opportunities for process improvement. Similar to planning activities, that are conducted at the beginning of the project, and at the beginning of each sprint, the reflection activities are conducted at the end of each sprint and also at the conclusion of the project.

SPRINT REVIEW

At the completion of each sprint, the product owner, team members including scrum master, customer representatives, and other appropriate participants, as it is seen fit by the product owner, are invited to participate in the sprint review meeting. Sprint review allows the product owner and customer to learn about the progress of the team, and for the team to learn about the goals and vision of the product owner and the customer for the future sprints. In the case of a software product, the participants in the sprint review work with the running software, therefore it is a hands-on activity by the product owner and customer/users rather than the demonstration of the software by the team. This meeting like other meetings in the scrum is time boxed (i.e., two hours for a two week sprint), and all participants are free to ask question and provide input to the meeting. The main purpose of the sprint review is to inspect the product, in order to make sure it is meeting the sprint goal and deliver the intended value to the customer. In another words, the sprint review serves as the validation of the requirements (backlog items) that has been selected to be delivered in the sprint. Sprint review also serves as an avenue for the customer to provide feedback about the delivered increment in the form of recommendations for potential improvement to the product (i.e., additional features). Of course, if any new functionality (backlog items) is identified, then these need to be added to the product backlog. Another activity in the sprint review meeting is the collaboration and discussion of the participants regarding the goals and objective of the project and the future delivery of the product. This information will provide important information to be considered during the future sprint planning. Overall, the sprint review provides the following advantages:

- Provides frequent engagement opportunities to the stakeholders (i.e., customer, product owner, and team)
- Provides visibility to the product quality, and progress
- Increases the cross team product understanding (for projects that have multiple teams)
- Provides information to be considered for the future sprint planning

SPRINT RETROSPECTIVE

At the completion of each sprint, the scrum master, and the rest of the development team will conduct a sprint retrospective meeting, with the purpose of reviewing the process used in the sprint, and identifying potential process improvement opportunities. This meeting like other scum meetings is time boxed, and it is typically 45 minutes for each week in the sprint, therefore the sprint retrospective for a two-week sprint is about one and half hour. It is the responsibility of the scrum master to provide a safe space for the team to openly discuss between themselves how the team performed during the just completed sprint and come up with the answers to the following questions. It is also important to address these questions in the same order that they are presented here. This allows the team to start the discussion by highlighting the positives ("our wins"), and then follow on by pointing out the issues, and finally identifying the opportunities for improvement.

- **What went well during the sprint?** Here we are looking to identify the processes, activities, techniques, tools, trainings, etc., that were used during the sprint, that resulted in the successful completion of the sprint. Once these are identified, we want to make sure we do the same during the following sprint.

- **What went wrong during the sprint?** Here we are looking to identify things that went wrong during the sprint. Try to identify the root cause of these problems, and find out when we first recognized them. Perhaps by identifying the root cause of the issues, and the time we first recognized them, we can implement corrective actions that will prevent us from doing the same in the future sprints.
- **What did we learn during the sprint?** This question is partly answered by the answer from the previous two questions. The first question will identify the processes, activities, techniques, etc., that have been useful, and it will also provide us with the information about the things that were not useful. The second question will provide us the information about the root cause of our problems. These are valuable lessons learned during the sprint, and provide us with the valuable information to carry on to the next sprint.
- **What do we need to do differently next sprint?** Now that we have identified the things that went well and the things that went wrong, and the corresponding lessons learned, we can discuss the things that can potentially be done differently in the future in order to improve the process.

It is important to make sure that there are some actions that take place as a result of the sprint retrospective, in order to show its value. To do this, at the completion of the sprint retrospective, the team selects one process improvement opportunity that they will attempt to implement during the next sprint. This is referred to as **Kaizen**, which is a Japanese term for a small incremental improvement over a long term. The idea here is that no matter how small the improvement is, sprint after sprint, we will get better over time.

The sprint review and retrospective provide the team an opportunity to reflect on the product that was completed in that sprint and the process that was used to do the work. The team also conducts the review and retrospective meeting to reflect on the product and processes that have been used throughout the project life cycle. While the reflection at the end of each sprint provides information that can be used tactically in order to improve the product and process for the future sprints, the reflection at the end of the project could potentially provide information that can be used strategically for the future projects, therefore getting better one project at the time.

Case Study Exercises

EXERCISE 10.1: FORMING THE DH SCRUM TEAM

SCENARIO

As part of the DH project in the previous chapters, a team has been formed to work on the development of the project following a plan-driven process. In this exercise, we plan to form a Scrum team that will develop the prototype.

LEARNING OBJECTIVES

Upon completion of this exercise, students will have the ability to:

- Identify the personnel that have the expertise to complete project.
- Identify the Scrum Master and the Product Owner for the project.

EXERCISE DESCRIPTION

1. This exercise follows a lecture on Scrum Process Overview.
2. As preparation for this exercise students are required to read the following:
   - DH Beginning Scenario
3. In this exercise, student groups will evaluate the existing team formation against the Scrum team characteristics. The exercise should include the following evaluation:
   - Do we have the right team members to form a Scrum team?
   - If so, which team members have the qualifications to serve as Product Owner and Scrum Master?
   - Do the remaining original team members have the appropriate qualification to be part of the team?
   - Is there anyone else that should be included as part of the Scrum organization?

EXERCISE 10.2A: DETERMINING DH BACKLOG ITEM

SCENARIO

After the completion of the tutorial on Scrum process, Massood asked the DH Team to establish an initial set of the backlog items for the Digital Home, given the DH Need Statement document.

LEARNING OBJECTIVES

Upon completion of this exercise, students will have increased ability to:

- Identify the product backlog items for a project.
- Ability to assign prioritize these backlog items.

Exercise Description

1. This exercise follows a lecture on Scrum process.
2. As preparation for this exercise students are required to read the following:
   - DH Beginning Scenario
   - *DigitalHome* Need Statement
3. Students are grouped into teams of 3–4.
4. Each team reviews and discusses the DH Need Statement.
5. Each team will generate an initial set of backlog items for the Digital Home.
6. Each team will prioritize these backlog items and the reasoning behind their prioritization.
7. The team presents its findings to the rest of the class.

EXERCISE 10.2B: BACKLOG ESTIMATION

Scenario

After the completion of the tutorial on Scrum process, Massood asked the DH Team to establish an initial set estimation for the backlog items that have been generated as a result of exercise 10.1 using two different estimation techniques (T-shirt sizing and playing poker).

Learning Objectives

Upon completion of this module, students will have increased ability to:

- Estimate backlog items using T-shirt sizing and playing poker.
- Be able to discuss the advantages and disadvantages of the two techniques.

Exercise Description

1. This exercise follows a lecture on estimation techniques.
2. The result of exercise 10.1 is needed as an input for this exercise.
3. Students are grouped into teams of 3–4.
4. Each team reviews and discusses the T-shirt sizing and playing poker estimation techniques.
5. Each team will use the T-shirt sizing technique to generate the estimate for the backlog Item.
6. Each team will use the playing poker technique to generate the estimate for the backlog item.
7. Each team compares the two estimation techniques and identifies the advantages and disadvantages of each technique.
8. Each team presents its finding to the class.

EXERCISE 10.3: SPRINT PLANNING

Scenario

After the completion of the tutorial on Scrum process, Massood asked the DH Team to identify all the backlog items that are addressing the Digital Home temperature control. Assuming the goal for the next sprint is to deliver the Digital Home temperature control functionality. Conduct the appropriate backlog grooming in order to generate the sprint backlog item.

Learning Objectives

Upon completion of this module, students will have increased ability to:

- Identifying the highest priority backlog items.
- Conduct backlog grooming.
- Generate the sprint backlog items.

Exercise Description

1. This exercise follows a lecture on Scrum process.
2. Students are grouped into teams of 3–4 people.
3. Each team will conduct backlog grooming for identification of the sprint backlog item to deliver the temperature control functionality.
4. Each team generates its own sprint backlog.
5. Each team reports their result to the class.

References

[ABET 2018] *Criteria for Accrediting Engineering Programs, 2019–2020 Accreditation Cycle*, Engineering Accreditation Commission, ABET Inc., November 2, 2018.

[ACM 1999] ACM/IEEE-CS Joint Task Force on Software Engineering Ethics and Professional Practices, *Software Engineering Code of Ethics and Professional Practice* (Version 5.2), 1999. (http://www.acm.org/serving/se/code.htm).

[ACM 2009] *Graduate Software Engineering 2009 (GSwE2009) Curriculum Guidelines for Graduate Degree Programs in Software Engineering*, Version 1.0, Integrated Software & Systems Engineering Curriculum (iSSEc) Project, ACM and IEEE-CS, 2009. (http://www.computer.org/portal/web/education/Curricula).

[ACM 2015] ACM/IEEE-CS Joint Task Force on Computing Curricula, *Software Engineering 2014, Curriculum Guidelines for Undergraduate Degree Programs in Software Engineering*, February 2015. (http://www.acm.org/education/se2014.pdf).

[Anderson 1993] Anderson, R. E. et al., Using the ACM Code of Ethics in Decision Making, *Communications of the ACM*, February 1993.

[Armour 1993] Armour, J. and Humphrey, W., *Software Product Liability*, Technical Report, CMU/SEI-93-TR-13, Software Engineering Institute, Carnegie Mellon University, August 1993.

[Basili 1984] Basili, V. and Weiss, D., A Method for Collecting Valid Software Engineering Data, *IEEE Transactions on Software Engineering*, November 1984.

[Bass 2012] Bass, L., Clements, and Kazman, R., *Software Architecture in Practice*, 3rd Edition, Boston, MA: Addison-Wesley, 2012.

[Beck 2000] Beck, K., *Extreme Programming Explained: Embrace Change*, Boston, MA: Addison-Wesley, 2000.

[BLS 2016] *Occupational Outlook Handbook 2016–17 edition*, Bureau of Labor Statistics, (http://www.bls.gov/oco/).

[Boehm 1988] Boehm, B., A Spiral Model of Software Development and Enhancement, *Computer*, Vol. 21, No. 5, pp. 61–72, May 1988.

[Boehm 2000] Boehm, B., et al., *Software Cost Estimation with Cocomo II*, Upper Saddle River, NJ, Prentice Hall, 2000.

[Boehm 2004] Boehm, B. and Turner, R., *Balancing Agility and Discipline*, Reading, MA: Addison-Wesley. 2004.

[Booch 1999] Booch, G., Rumbaugh, J., and Jacobson, I., *The Unified Modeling Language User Guide*, Reading, MA: Addison-Wesley, 1999.

[Booch 2004] Booch, G., *Object-Oriented Analysis and Design with Applications*, 3rd Edition, Reading, MA: Addison Wesley, 2004.

[Bourque 2014] Bourque, P. and Fairley, R., Eds., *Guide to the Software Engineering Body of Knowledge*, Version 3, Los Alamitos, CA: IEEE Computer Society, 2014.

[Brooks 1995] Brooks, F. P., *The Mythical Man-Month, 20th Anniversary Edition*, Reading, MA: Addison Wesley, 1995.

[Budgen 1989] Budgen, D., *Introduction to Software Design*, SEI, Carnegie Mellon University Technical Report, Curriculum Module SEI-CM-2-2.1, January 1989.

[Chidamber 1994] Chidamber, S. R. and Kemerer, C. F., A Metric Suite for Object Oriented Design, *IEEE Transactions on Software Engineering*, Vol. 20, pp. 476–493, 1994.

[Clements 2010] Clements, C. et al., *Documenting Software Architectures: Views and Beyond*, 2nd Edition. Reading, MA: Addison-Wesley, 2010.

[Davis 2003] Davis, N. and Mullaney, J., *The Team Software Process (TSP) in Practice: A Summary of Recent Results*, CMU/SEI-2003-TR-014, Software Engineering Institute, Carnegie Mellon University, September 2003.

[Dehaghani 2013] Dehaghani, S. and Hajrahimi, N., Which Factors Affect Software Projects, Maintenance Cost More? *Acta Informatica Medica*, Vol. 21, No. 1, pp. 63–66, March 2013.

[DeMarco 1982] DeMarco, T., *Controlling Software Projects*, New York: Yourdon Press, 1982.

[DeMarco 1999] DeMarco, T., and Lister, T., *Peopleware: Productive Project and Teams*, 2nd Edition, New York: Dorset House, 1999.

[Fagan 1986] Fagan, M., Advances in Software Inspection, *IEEE Transactions in Software Engineering*, Vol. 12, No. 7, pp. 744–751, July 1986.

[Fairley 2009] Fairley, R. E., *Managing and Leading Software Projects*, Hoboken, NJ: Wiley-IEEE Computer Society Press, 2009.

[Ford 1996] Ford, G. and Gibbs, N. E., *A Mature Profession of Software Engineering*, CMU/SEI-96-TR-004, Software Engineering Institute, Carnegie Mellon University, January 1996.

[Fowler 2004] Fowler, M., *UML Distilled: A Brief Guide to the Standard Object Modeling Language*, 2nd Edition, Reading, MA: Addison Wesley, 2004.

[Gamma 1995] Gamma, E., Helm, R., Johnson, R., and Vlissides, J., *Design Patterns: Elements of Reusable Object-Oriented Software*, Reading, MA: Addison-Wesley, 1995.

[Ganis 2007] Ganis, M., Creating Effective Teams (Without having to herd cats), *2007 New York Library Association Annual Conference*, October 17–20, 2007. (http://webpage.pace.edu/mganis/nyla.htm).

[Herreid 1994] Herreid, C., Case Studies in Science: A Novel Method of Science Education, *Journal of College Science Teaching*, Vol. 23, pp. 221–229, February 1994.

[Highsmith 2001] Highsmith, J. and Cockburn, A., Agile Software Development: The Business of Innovation, *Computer*, Vol. 34, pp. 120–122, September 2001.

[Hofmeister 2000] Hofmeister, C., Nord, R., and Soni, D., *Applied Software Architecture*, Boston, MA: Addison-Wesley, 2000.

[Humphrey 1995] Humphrey, W. S., *A Discipline for Software Engineering*, Reading, MA: Addison Wesley, 1995.

[Humphrey 2000] Humphrey, W. S., *Introduction to the Team Software Process*, Reading, MA: Addison-Wesley, 2000.

[IEEE 730] IEEE Std 730-2014, *IEEE Standard for Software Quality Assurance Processes*, IEEE Computer Society, 2014.

[IEEE 830] IEEE Std 830-1998, *IEEE Recommended Practice for Software Requirements Specifications*, IEEE Computer Society, 1998.

[IEEE 1012] IEEE Std 1012-2012, *IEEE Standard for System and Software Verification and Validation*, IEEE Computer Society, 2012.

[IEEE 1016] IEEE Std 1016-2009, *IEEE Standard for Information Technology - Systems Design - Software Design Descriptions*, IEEE Computer Society, 2009.

[IEEE 1028] IEEE Std 1028-2008, *IEEE Standard for Software Reviews and Audits*, IEEE Computer Society, 2008.

[IEEE 1061] IEEE Std 1061-1998, *IEEE Standard for a Software Quality Metrics Methodology*, IEEE Computer Society, 1998.

[IEEE 12207] IEEE Std 12207-2008, *Systems and Software Engineering – Software Life Cycle Processes*, IEEE Computer Society, 2008.

[IEEE 14764] IEEE Std 14764-2006, *Software Engineering-System Life Cycle Processes – Maintenance*, IEEE Computer Society, 2006.

[IEEE 24765] IEEE Std 24765:2010, *IEEE Standard Systems and Software Engineering Vocabulary*, IEEE Computer Society, 2010.

[IEEE 29148] IEEE Std 29148, *IEEE Standard for Systems and Software Engineering – Life Cycle Processes — Requirements Engineering*, IEEE Computer Society, 2011.

[Jones 2003] Jones, C., *Variations in Software Development Practices*, IEEE Software, November/December 2003.

[Kan 2002] Kan, S. H., *Metrics and Models in Software Quality Engineering*, 2nd Edition, Boston, MA: Addison-Wesley, 2002.

[Kruchten 2004] Kruchten, P., *The Rational Unified Process: An Introduction*, 3rd Edition, Reading, MA: Addison-Wesley, 2004.

[Lientz 1981] Lientz, B. P. and Burton Swanson, E. B., Problems in Application Software Maintenance, *Communications of the ACM*, Vol. 24, No. 11, pp. 763–769, 1981.

[Marandi 2014] Marandi1, A. K. and Khan, D. A., Analytical Phase Wise Analysis of Defect Removal Effectiveness to Enhancing the Software Quality, *International Proceedings of Economics Development and Research*, Vol. 75, pp. 40–46, January 2014.

[McCabe 1976] McCabe, T. J., A Complexity Measure, *IEEE Transactions on Software Engineering*, Vol. SE-2, pp. 308–320, 1976.

[McConnell 2004] McConnell, S., *Code Complete: A Practical Handbook of Software Construction*, 2nd Edition, Redmond, WA: Microsoft Press, 2004.

[Moore 2006] Moore, J. W., *The Road Map to Software Engineering: A Standards-Based Guide*, Hoboken, NJ: Wiley-IEEE Computer Society Press, 2006.

[Naur 1969] Naur, P. and Randell, B., Eds., *Software Engineering: Report on the NATO Software Engineering Conference 1968*, NATO Science Committee, January 1969.

[Oakley 2004] Oakley, B., Felder, R., Brent, R., and Elhajj, I., Turning Student Groups into Effective Teams, *Journal of Student Centered Learning*, Vol. 2, No. 1, pp. 9–34, 2004.

[Parnas 1972] Parnas, D. L., On the Criteria to be Used in Decomposing Systems in Modules, *Communications of the ACM*, Vol. 15, No. 12, pp. 1053–1058, December 1972.

[Parnas 2003] Parnas, D. L. and Lawford, M., The Role of Inspection in Software Quality Assurance, *IEEE Transactions on Software Engineering*, Vol. 29, No. 8, pp. 674–676, August 2003.

[Paulk 1993] Paulk, M. C. et al., *Capability Maturity Model for Software*, Version 1.1, CMU/SEI-93-TR-024, Software Engineering Institute, Carnegie Mellon University, 1993.

[Paulk 2001] Paulk, M. C., Extreme Programming from a CMM Perspective, *IEEE Software*, Vol. 18, No. 2, pp. 19–26, November/December 2001.

[Pfleeger 2009] Pfleeger, S. and Atlee, J., *Software Engineering Theory and Practice*, 4th Edition, Englewood Cliffs, NJ: Prentice-Hall, 2009.

[Pressman 2015] Pressman, R. and Maxim, B., *Software Engineering: A Practitioner's Approach*, 8th Edition, New York: McGraw-Hill, 2015.

[Royce 1970] Royce, W., Managing the Development of Large Software Systems, *Proceedings of IEEE WESCON 26*, pp 1–9, August 1970.

[Schwaber 2004] Schwaber, K., *Agile Project Management with Scrum*, Redmond, WA: Microsoft Press, 2004.

[Schwaber 2017] Schwaber, K. and Sutherland, J., *The Scrum Guide*, (https://www.scrumguides.org/docs/scrumguide/v2017/2017-Scrum-Guide-US.pdf).

[Scott 2002] Scott, K., *The Unified Process Explained*, Reading, MA: Addison-Wesley, 2002.

[SEI 2010] *CMMI for Development*, Version 1.3, CMU/SEI-2010-TR-033, Software Engineering Institute, Carnegie Mellon University, November 2010. (http://www.sei.cmu.edu/reports/10tr033.pdf).

[Shaw 1996] Shaw, M. and Garlan, D., *Software Architecture: Perspectives on an Emerging Discipline*, Upper Saddle River, NJ: Prentice-Hall, 1996.

[Sommerville 2011] Sommerville, I., *Software Engineering*, 9th Edition. New York: Addison-Wesley, 2011.

[Standish 2020] The Standish Group International, Inc., *CHAOS Report 2020*, January 2020. (https://www.standishgroup.com).

[Stevens 1974] Stevens, W., Myers, G., and Constantine, L., Structured Design, *IBM Systems Journal*, Vol. 13, No. 2, pp. 115–139, 1974.

[Sun 1997] *Java Code Conventions*, Sun Microsystems, Inc., September 12, 1997.

[Sutherland 2019] Sutherland, J., The Scrum@Scale Guide, November 2019. (https://www.scrumatscale.com/wp-content/uploads/Scrum@Scale-Guide.pdf).

[SWECOM 2014] *Software Engineering Competency Model (SWECOM)*, Version 1.0, IEEE Computer Society, 2014.

[Tassey 2002] Tassey, G., *The Economic Impacts of Inadequate Infrastructure for Software Testing*, Washington, D.C.: National Institute of Standards and Technology, U.S Department of Commerce Technology Administration, May 2002.

[Weiss 1990] Weiss, E. A., Ed., Self-Assessment Procedure XXII, *Communications of the ACM*, November 1990.

[Wiegers 2003] Wiegers, K. E., *Software Requirements*, 2nd Edition. Redmond, WA: Microsoft Press, 2003.

Appendix A: Digital Home Customer Need Statement (9/12/202X)

This is a customer need statement for the development of a "Smart House", called "Digital Home", prepared by the Marketing Division of *HomeOwner, Inc*, for its *DigitalHomeOwner* Division. A "Smart House" is a home with a digital management system that allows home owners (or renters) to easily manage their daily lives by providing for a lifestyle that brings together home security, environmental, and energy management (temperature, humidity, and lighting), entertainment, and communications. The Smart House components consist of household devices (e.g., an air conditioning unit, a sound system, a water sprinkler system, a home security system, etc.), sensors and controllers for the devices, communication links between the components, and a computer system, which will manage the components.

The need statement will be used by *DigitalHomeOwner* to develop a prototype "Digital Home" system to assess the technical and business issues associated with such development. Although the statement about the Wright family is fiction, it represents information and opinion collected from a variety of potential users and other sources, and provides a realistic basis for development of the Digital Home prototype.

Customer Profile: The Wright Family

Steve Wright – 76 years old – Retired Documentary Movie Director – disabled
Mini Wright – 72 years – Housewife
Stanley Wright – 47 years – Son and Product Manager of Splice, Inc
Vinni Wright – 42 years – Housewife and an online Library Science student at Cal State, Los Angeles

Michelle Wright – 18 years – freshman college student (Bachelor of Arts student) at Cal State, Los Angeles
Robert Wright – 16 years – high school student
All live as a joint family in Los Angeles, CA.

Introduction

Steve and Stanley Wright have decided to buy a new house in Los Angeles, CA. Since they are a well-off family, they have decided to acquire a house where all the members can live comfortably, efficiently, and safely. With a large, modern house comes a greater desire to ensure that the house is well maintained and all the facilities are easy to access. So, the Wrights are willing to spend about 10% more than current house value (up to $50,000 additional), to add some automated facilities to the house. They want a house that is easy to maintain and also has a two-year warranty on the structure and its infrastructure. They also want the house warranty to be renewable (just like a new car).

Many things have influenced the Wrights' housing needs and desires; one of them being that Steve's wife, Mini, has a heart condition. He wants to make sure that his wife is well-monitored and taken care of as soon as she has any trouble. Another reason to buy such a house is the need to provide Steve good mobility and accessibility in the house, since he has to use a wheelchair. Another motivation is that Stanley's wife, Vinni, is a very busy housewife, managing a large family and taking online courses. The children are typical 21st century teenagers and want the latest technology available in the house. The Wrights also believe in conserving energy and have been a part of many energy conservation programs such as SaveEarth (http://www.save_earth.org). Also, in this multi-generational family, support for effective communication and security are essential.

Super Contractors, Inc, decided to take on the Wright home as one of their major projects on their 50th anniversary of home building. To support their efforts they have engaged the *DigitalHomeOwner* Division of *HomeOwner*, Inc, to assist in the development and implementation of the "smart" features of the new house. Super Contractors has set up interview of the family, with *DigitalHomeOwner* personnel participating. The following section is based on the interview.

Wright Family Needs and Desires

The family members are looking for a wireless, safe and secure, and an easily usable house with good home entertainment facilities and other facilities that bring a convenient and efficient lifestyle to the Wright family.

For Steve, the most important aspect in the house is accessibility. He wants to be able to control the facilities in the house with voice recognition. This will make his life simpler since he can control the TV, the music, the lighting, and the air with his voice (at a minimum, in his room). To enter the house, he wants the house door to open automatically with a code.

Mini has heart problems and wants to make sure that the house accommodates her special health care needs. She wants a heart monitoring system that will assess her heart rate and blood pressure every morning, and is capable of notifying emergency services in case she has an attack. Because of her condition, she is also concerned with the quality of air that she and her family breathe. She wants air purification for the entire house and also a way of determining the air quality (e.g., humidity, temperature, carbon monoxide, smoke, fire, dust, and pollen levels).

For Stanley, energy conservation is a very important factor. He wants the lights to be capable of turning on and off automatically, depending on where the family members are located. He also wants solar panels that are capable of helping with energy conservation by providing heated water and electricity. Since power outages may occur, he also wants to control the status of the electronic appliances in case of emergency and have a backup power source that may take over, and re-establish power for the house.

Vinni is a busy housewife since she has to cook and clean every day. However, she is also taking online courses for a Library Science degree. Although the rest the family helps out, she needs additional assistance with the daily household chores. She wants to be able to see the status of various household devices. For example, if she puts clothes in the dryer, and she is standing in the kitchen cooking dinner, she wants a message on the fridge telling her that the dryer cycle is complete. She also wants to be able to clean the kitchen and vacuum the house using automatic features. When she is cooking, she would like independent and automatic control of all stove heating elements. When the dishwasher has a certain number of dishes, it could be set to start dishwashing automatically.

The family would like the capability to manage their appliances and automated features through touch screens located in various rooms that permit them to interact and control all of the features at the touch of a button/screen. The interface should save time and increase the control, communication, and livability within their home. As an option, they also want to be able to control their home devices through the Internet, allowing communication and control even when not at home. For example, they would like to be able to adjust the lights, set the alarm system, adjust the A/C temperature, and verify that windows and doors are shut, all over the internet.

Michelle is a fun loving teenager with a "cool" attitude and a bunch of friends who come and party at her place. She wants to have the place ready and nice for her social gatherings. She is interested in the idea of entertainment modes (date mode, pool party, etc.) which enable certain desired features. For example, when she selects pool party mode, she wants a good selection of party music, along with the correct lighting that sets the mood for the party. She also wants her entertainment system to be capable of recording her favorite shows while she is away at school.

Robert is a "high-tech" teenager, with all the latest techie devices (tablet, etc.). He would like to have the ultimate in technically advanced house; besides wanting intelligent appliances that "interact with humans", he is also interested on keeping up with technology and having the latest gadgets and tools. He wants a house that is capable of knowing when he is close to home after school and has everything ready (e.g., find the TV channels and programs he likes on TV, set A/C temperature in his room, prepare his favorite snack, etc.)

The Wright family believes in living in a "home" and not a house with a lot of automation and digitization, giving it the look and feel of a laboratory. They want to feel at home when they are in this automated house. If that involves removing the voice activated system and using only touch pad system, then they are willing to take that. Most of the family members have a high level of knowledge in technology and use of software; but, Mini still prefers older, traditional ways of dealing with household features. She wants to make sure that all the technology is easy to use and understand.

The whole family wants a safe and reliable house that takes good care of the members. However, the major concern of Stanley (since he has a good background in this area) is that the house will not start controlling the people in the house. Basically, he does not want the house to tell them what to do by learning their habits. For example, the house should not be putting ingredients into their food! Or it should not be taking care of the children in the house. The whole family wants to have manual control of the house at any point or time during their use to ensure their safety in case of any system failure.

The family also looks for a secure place to live. Therefore, they have concerns about acquiring an automated home security system which includes alarm and sensors that are easily manageable and can notify the police and fire department in case of an accident. The automated system should also turn on the lights if any intruders come in, and notify the owner and police department if someone actually does break in. The system should also prevent unauthorized personnel from gaining access to or controlling the automatic features of the house.

Appendix B: *DigitalHome* Software Requirements Specification

Version 1.2
   October 18, 202X

Change History

*Added (A), Modified (M), Deleted (D)

Version	Date	AMD*	Author	Description	Change Request #
1.0	10/11/202X	A	Michel Jackson	First Version	
1.1	10/15/202X	AM	Michel Jackson	Updated version, based on external review	1,2,3, 4, 5, 6, 7

(Continued)

Version	Date	AMD*	Author	Description	Change Request #
1.2	10/18/202X	AM	Michel Jackson	Updated version, based on management requests	8, 9

Contents

DigitalHome Requirements Specification 289
Introduction 289
Team Project Information 289
Overall description 290
    Product Description and Scope 290
    Users Description 290
    Development Constraints 291
    Operational Environment 292
Functional Requirements 293
    General Requirements 293
    Thermostat Requirements 294
    Humidistat Requirements 294
    Appliance Management Requirements 295
    DH Planner Requirements 295
    DH Planner Requirements 295
Other Non-Functional Requirements 296
    Performance Requirements 296
    Reliability 296
    Safety Requirements 297
    Security Requirements 297
    Maintenance Requirements 297
    Business Rules 297
    User Documentation: 298
References 298
Appendix – Use Case Diagram 299

DigitalHome Requirements Specification

1. Introduction

This document specifies the requirements for the development of a "Smart House", called *DigitalHome* (DH), by the *DigitalHomeOwner* Division of *HomeOwner*, Inc. A "Smart House" is a home management system that allows home owners (or renters) to easily manage their daily lives by providing for a lifestyle that brings together home security, environmental and energy management (temperature, humidity and lighting), entertainment, and communications. The Smart House components consist of household devices (e.g., a power and lighting system, a heating and air conditioning unit, a home security system, a sound system, a water sprinkler system, small appliances, etc.), sensors and controllers for the devices, communication links between the components, and a computer system, which will manage the components.

The *DigitalHome* Software Requirements Specification (SRS) is based on the *DigitalHome Customer Need Statement*. It is made up of a list of the principle features of the system. This initial version of *DigitalHome* will be a limited prototype version, which will be used by *HomeOwner* management to make business decisions about the future commercial development of *DigitalHomeOwner* products and services. Hence, the SRS is not intended as a comprehensive or complete specification of *DigitalHome* requirements. This document was prepared by the *DigitalHomeOwner* Division, in consultation with the Marketing Division of HomeOwner, Inc.

2. Team Project Information

- Members/Roles
  - Team Leader: Disha Chandra
  - System Analyst and Requirements Manager: Michel Jackson
  - System Architect and Design Manager: Yao Wang
  - Planning Manager: Georgia Magee
  - Quality Manager: Massood Zewail
  - Construction Engineer: all team members
- Schedule
  - Need Assessment – 9/15/202X
  - Project Launch – 9/16/202X
  - Project Plan – 9/27/202X
  - Requirements – 10/18/202X
  - Architecture -11/15/202X
  - Cycle 1 Construction – 12/15/202X
  - Cycle 2 Construction – 1/17/202Y
  - Cycle 3 Construction –2/14/202Y
  - Cycle 4 Construction –3/24/202Y

- Cycle 5 Construction – 4/18/202Y
- Cycle 6 Construction – 5/16/202Y
- System Testing – 6/20/202Y
- Acceptance Testing – 7/14/202Y
- Postmortem Analysis – 8/1/202Y
- System Testing – 6/20/202Y
- Acceptance Testing – 7/14/202Y
- Postmortem Analysis – 8/1/202Y

3. Overall description

3.1 Product Description and Scope

The Digital Home system, for the purposes of this document, is a system that will allow a home user to manage devices that control the environment of a home. The user communicates through a personal web page on the *DigitalHome* web server. The DH web server communicates, through a home wireless gateway device, with the sensor and controller devices in the home. The product to be produced will be a prototype which will allow business decisions to be made about future development of a commercial product. The scope of the project will be limited to the management of devices which control temperature, humidity and power to small appliances, through the use of a web-ready device.

3.2 Users Description

3.2.1 DigitalHome Users

3.2.1.1 A user shall be able to use the DH system capabilities to monitor and control the environment in his/her home.

3.2.1.2 A user is familiar with the layout of his/her home and the location of sensor and control devices (for temperature, for humidity, and for power to small appliances).

3.2.1.3 Although a user may not be familiar with the technical features of the DH system, he/she is familiar with the use of a web interface and simple web operations (logging in and logging out, browsing web pages, and submitting information and requests via a web interface).

3.2.1.4 A Master User will be designated, who will be able to manage user profiles and to limit or constrain the use of the system for other users. For example, the Master User can add new users, delete existing users, or may restrict who may or may not control the security system.

3.2.2 DigitalHome Technician

3.2.2.1 A DH Technician is responsible for setting up and maintaining the configuration of a DH system.

3.2.2.2 A DH Technician has experience with the type of hardware, software, and web services associated with a system like the DH system.

3.2.2.3 A DH Technician is specially trained by DigitalHomeOwner to be familiar with the functionality, architecture, and operation of the DH system product.

3.2.2.4 A DH Technician also has the same rights as a DH Master user.

3.3 Development Constraints

3.3.1 The "prototype" version of the DigitalHome System (as specified in this document) must be completed within twelve months of inception.

3.3.2 The development team will consist of five engineers. The DigitalHomeOwner Director will provide management and communication support.

3.3.3 The development team will use the Team Software Process (TSP) [Humphrey 2000] for this project.

3.3.4 Where possible, the DigitalHome project will employ widely used, accepted, and available hardware and software technology and standards, both for product elements and for development tools. See section 3.4 for additional detail.

3.3.5 Because of potential market competition for DigitalHome products, the cost of DigitalHome elements for this project should be minimized.

3.3.6 The DH system will be tested in a simulated environment. There will be no actual physical home and there will be a mixture of actual physical sensors and controllers, and simulated sensors and controllers. However, the simulated environment will be realistic and adhere to the physical properties and constraints of and actual home and to real sensors and controllers.

3.3.7 Changes to this document (e.g., changes in requirements/must be approved by the Director of the DigitalHomeOwner Division.

3.3.8 The Development Team will implement a final acceptance test process involving potential users and the upper level management of HomeOwner.

3.4 Operational Environment

Although the system to be developed is a "proof of concept" system intended to help *HomeOwner*, Inc, to make marketing and development decisions, the following sections describe operational environment concerns and constraints; some of them are related to issues of long-term production and marketing of a *DigitalHome* product.

3.4.1 The home system shall require and Internet Service Provider (ISP). The ISP should be widely available (cable modem, high speed DSL), such as Bright House or Bellsouth FastAccess.

3.4.2 DigitalHomeOwner DH Web Server

3.4.2.1 The DigitalHome Web server shall provide the capability for establishing and maintaining DH User Accounts and for interaction between a DH user and a DH user account.

3.4.3 Home DH Gateway Device

3.4.3.1 The DH Gateway device shall provide communication with all the DgitalHome devices and shall connect with a broadband Internet connection.

3.4.3.2 The Gateway contains an RF Module, which shall send and receive wireless communications between the Gateway and the other DigitalHome devices (sensors and controllers). The device operates at 418 MHz with up to a 100-foot range for indoor transmission.

3.4.4 Sensors and Controllers

3.4.4.1 The system includes digital programmable thermostats, which shall be used to monitor and regulate the temperature of an enclosed space. The thermostat provides a reading of the current temperature in the space where the thermostat is located; and provides a "set point" temperature that is used to control the flow of heat energy (by switching heating or cooling devices on or off as needed) to achieve the set point temperature. The sensor part of the thermostat has a sensitivity range between 14°F and 104°F (–10°C and 40°C).

3.4.4.2 The system includes digital programmable humidistats, which shall be used to monitor and regulate the humidity of an enclosed space. The humidistat provides a reading of the current humidity in the space where the humidistat is located; and provides a "set point" humidity that is used to control humidifiers and dehumidifiers achieve the set point humidity.

3.4.4.3 The system includes magnetic alarm contact switches which are used monitor improper entry through a door or window.

3.4.4.4 The system includes home security alarms and lights, which can be activated by when DigitalHome senses a home security breach from a magnetic contact.

3.4.4.5 The system includes digital programmable power switches which shall be used to monitor the current state of an appliance (e.g., a lamp or a coffee maker) or to change the state of the appliance.

3.4.4.6 The system includes digital light sensors, which shall measure the amount of light that they sense. The light sensor reports the amount of light as a number between 0 (total darkness) and 100 (very bright).

4. Functional Requirements

This section provides a description of the functional requirements. There is use case diagram in the appendix, which provides an overview of the system functionality and shows the relationships between the *DigitalHome* System entities.

4.1 General Requirements

4.1.1 The DigitalHome System shall allow any web-ready computer, cell phone or PDA to control a home's temperature, humidity, lights, home security, and the state of small appliances. The communication center of the system shall be a personal home owner web page (maintained by DigitalHomeOwner - at http://www.DigitalHomeOwner.com), through which a user shall be able to monitor and control home devices and systems.

4.1.2 Each DigitalHome shall contain a master control device (the DH Gateway Device) that connects to the home's broadband Internet connection and uses wireless communication to send and receive communication between the DigitalHome system and the home devices and systems.

4.1.3 The DigitalHome shall be equipped with various environmental controllers and sensors (temperature controller-sensors, humidity controller-sensors, and power switches). Using wireless communication, sensor values can read and saved in the home database and values can be sent to controllers to change the environment.

4.2 Thermostat Requirements

4.2.1 The DigitalHome programmable thermostat shall allow a user to monitor and control a home's temperature from any place, using almost any web ready computer, cell phone, or PDA.

4.2.1.1 The DigitalHome system shall be able to set the thermostat temperatures to between 60° F and 80° F.

4.2.1.2 The DigitalHome system shall be able to read and display the temperature at a thermostat position.

4.2.2 Up to eight thermostats shall be placed throughout the home and shall be controlled individually or collectively, so that temperature can be controlled at different levels in different home spaces.

4.2.2.1 A single thermostat shall be placed in an enclosed space (e.g., a room in the house) for which the air temperature is to be controlled.

4.2.2.2 For each thermostat, up to four different settings per day for every day of the week can be scheduled.

4.2.2.3 Scheduled settings may be overridden at any time, either remotely or manually at the thermostat itself.

4.2.3 A thermostat unit shall communicate, through wireless signals, with the master control unit.

4.2.4 The system shall support Fahrenheit and Celsius temperature values.

4.2.5 The system shall compatible with most centralized HVAC (Heating, Ventilation and Air Conditioning) systems: gas, oil, electricity, solar, or a combination of two or more.

4.3 Humidistat Requirements

4.3.1 The DigitalHome programmable humidistat shall allow a user to monitor and control a home's humidity from any place, using almost any web ready computer, cell phone, or PDA.

4.3.1.1 A humidistat shall be able to humidify or dehumidify a living space.

4.3.2 Humidistats may be placed throughout the home and shall be controlled individually or collectively, so that humidity can be controlled at different levels in different home spaces.

4.3.3 A DigitalHome system shall use wireless signals to communicate, through the master control unit, with the humidistats.

4.3.4 A DigitalHome system shall be able to manage up to eight humidistats.

4.3.5 The user shall be able to select the humidity levels found most comfortable — from 30% to 60%.

4.4 Home Security System Requirements

4.4.1 The DigitalHome home security system consists of a set of contact sensors and a set of home security alarms, which are activated when there is a home security breach.

4.4.2 A DigitalHome system shall use wireless signals to communicate, through the master control unit, with sensors and alarms.

4.4.3 A DigitalHome system shall be able to manage up to thirty door and window sensors.

4.4.4 A DigitalHome system shall be able to activate both light and sound alarms.

4.4.5

4.5 Appliance Management Requirements

4.5.1 The DigitalHome programmable Appliance Manager shall provide for management of a home's small appliances and shall allow the user to turn lights on or off as desired.

4.5.2 The unit shall be able to control up to forty power switches, which control the on/off condition of lighting devices in a room and 115 volt, 10 amp appliances that plug into a standard wall outlet.

4.5.3 The appliance manager shall have access to up to eight light sensors.

4.5.4 The system shall be able to provide information about whether a device is off/on, or information about the illumination level in a room.

4.6 DH Planner Requirements

4.6.1 The DigitalHome Planner shall be able to provide a user with the capability to direct the system to set various home parameters (temperature, humidity, lighting, home security contacts, and on/off appliance status) for certain time periods.

4.6.2 DigitalHome provides a monthly planner on its website.

4.6.3 Parameter values can be scheduled on a daily or hourly basis.

4.6.4 All planned parameter values can be overridden by a user.

4.6.5 Various plan profiles (normal monthly profile, vacation profile, summer profile, Holiday profile, etc.) may be stored and retrieved to assist in planning.

4.6.6 The DigitalHome Planner shall be able to provide various reports on the management and control of the home (e.g., historical data on temperature, humidity, lighting, etc.).

5. Other Non-Functional Requirements

5.1 Performance Requirements

5.1.1 The DigitalHome web server system shall be able handle up to 150 users and up to 90 requests per second.

5.1.2 Displays of environmental conditions (temperature, humidity, power, and light) shall be updated at least every two seconds.

5.1.3 Sensor (temperature, humidity, power and light) shall have a minimum data acquisition rate of 10 Hz.

5.1.4 An environmental sensor or controller device shall have to be within 100 feet of the master control device, in order to be communication with the system.

5.2 Reliability

5.2.1 The DigitalHome System must be highly reliable with no more than 1 failure per 10,000 operations.

5.2.2 Long range plans for the DigitalHome system include a backup DH server to be used in place of the main DH server, in case of main server failure. The prototype version of DigitalHome shall provide for future addition of a backup server.

5.2.3 The Digital Home System shall incorporate adequate backup and recovery mechanisms in order to insure user planning and usage data is not lost due to system failure.

5.2.4 All DigitalHome operations shall incorporate appropriate exception handling so that the system responds to a user with a clear, descriptive message when an error or an exceptional condition occurs.

5.3 Safety Requirements

5.3.1 Although there are no specific safety requirements, high system reliability is important to insure are there are no system failures in carrying out user requests. Such failures might affect the safety of home dwellers (e.g., inadequate lighting in dark spaces, inappropriate temperature and humidity for people who are in ill-health, or powering certain appliances when young children are present).

5.4 Security Requirements

5.4.1 Upon installation, a DigitalHome user account shall be established. The DigitalHome web system shall provide for authentication and information encryption through a reliable and effective security technology (e.g., Transport Layer Security).

5.4.2 Log in to an account shall require entry of an account name and a password.

5.5 Maintenance Requirements

5.5.1 The DH system shall be easy to modify and add features.

5.5.2 Although the product produced under this document will be "prototype" version, all modules and components of this prototype version shall be designed and implemented in such a manner that they may be easily incorporated in a fully specified commercial version of the DigitalHome System.

5.5.3 The design and development of DH modules shall use abstraction and information hiding to insure high module cohesion and loose coupling between modules.

5.6 Business Rules

5.6.1 All system documents (Software Requirements Specification, Architectural Design Specification, Module Detailed Design, Module Source Code, and all Test Plans) shall be up-to-date, use the Homeowner document format [HO2305] and reside in the HomeOwner Document Archive at completion of the project.

5.6.2 HomeOwner has designated Java as the preferred language for development of software for HomeOwner products. Exceptions to this rule must be approved by the CIO.

5.7 Documentation Requirements

5.7.1 The DigitalHome System shall provide users with online documentation about the DigitalHome system installed in their home. The user documentation shall include the following:

5.7.1.1 An FAQ section – a set of "Frequently Asked Questions" about use and maintenance of the DigitalHome System (e.g., "How do I change my password?", "Where do I go to get DigitalHome support?").

5.7.1.2 A section that explains how DH parameters are set and sensor values are read. This shall include information on limitations and constraints on parameter settings and sensor reading accuracy.

5.7.1.3 A section on how to use of the DH Planner.

5.7.2 Since this version of DigitalHome is a prototype, which will be used by HomeOwner management to make business decisions about the future commercial development of DigitalHomeOwner products, it is essential that a complete, mature set of software artifacts be developed so that they can be used and reused for future development. In addition to a Software Requirements Specification, documentation should include the following artifacts:

■ Development Process
■ Project Plan (including a Task Plan and Schedule, a Risk Management Plan, and a Configuration Management Plan)
■ Software Design Description
■ Source Code
■ Unit Test Plans and Test Results Report
■ Integration Plan
■ System T System Test Plan and Test Results Report
■ Acceptance Test Results Report
■ Postmortem Report

6. References

[Humphrey 2000] Humphrey, W. S., *Introduction to the Team Software Process*, Addison-Wesley, 2000.

[HO2305] Sykes, J., and Rook, R., *Guidelines for Developing HomeOwner Technical Reports*, HO2305, March 2002.

[Meyer 2004] Meyer, G., *Smart Home Hacks: Tips & Tools for Automating Your House*, O'Reilly, 2004.

Appendix – Use Case Diagram

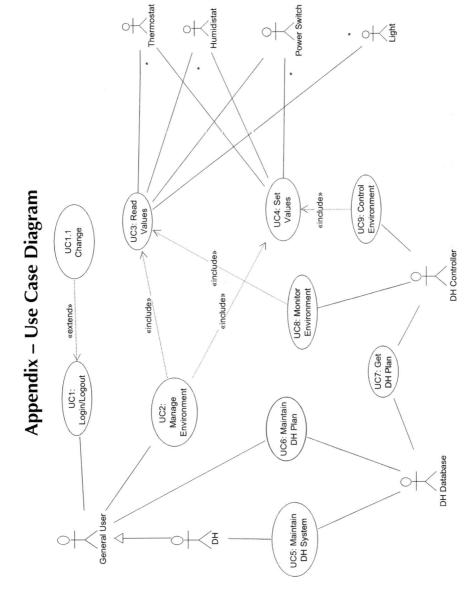

Appendix C: *DigitalHome* Use Case Model

HomeOwner Inc.

DigitalHomeOwner Division

Version 1.0
October 8, 202X

Contents

Introduction 302
DH Actors 302
Digital Home Use Case Diagram 303
   UC1 Use Case Diagram 304
   UC3 Use Case Diagram 305
UC1: Configure DH System..305
   UC1.1: Manage User Accounts 307
   UC1.2: Set Default Parameters.. 308
   UC1.3: Change DH System Operation State 310
UC2: Login/Logout... 311
UC3: Manage Monthly Planner 312
   UC3.1: Create Month Plan 313
      UC3.1.1: Set Temperature Plan Parameters......................... 315
      UC3.1.2: Set Humidistat Plan Parameters 316
      UC3.1.3: Set Security Plan Parameters.................................318
      UC3.1.4: Set Power Switch Plan Parameters 319
   UC3.2: Modify Month Plan 321
   UC3.3: View Month Plan..322

UC4: Prepare Month Report 323
UC5: Monitor Sensor Values 325
UC6: Read Sensor Values 326
UC7: Control DH Environment..327
UC8: Set Parameters Manually..329

Introduction

This document is a supplement to the *Digital Home Software Requirements Specification*, Version 1.3, developed by the *DigitalHomeOwner* Division of *HomeOwner*, Inc. It contains a Use Case Model that was developed as part of an analysis of the *DigitalHome* (DH) functional requirements. It includes a Use Case Diagram for the entire Digital Home system, along with several diagrams for specific parts of the system. For each use case, there is a scenario description that describes how actors interact with the system to carry out the use case functionality.

DH Actors

The use case diagrams and use case scenarios depict actors that interact with the *DigitalHome* System. The following is a description of the DH actors:

- *General User* – an actor who is able to use the DH system capabilities to monitor and control the environment in his/her home.
- *Master User* – a general user who has the ability to change the configuration of the DH System (e.g., add/delete users, change the default settings.
- *DH Technician* – a general user who is responsible for setting up and maintaining the configuration of a DH system.
- *DH Database* – a storage device that can be used to write or read user account information, monthly plan data, and default values.
- *DH Gateway* – a device, which provides communication between all the DH environment devices (sensors and controllers) the DH system.
- *Time* – A virtual actor that provides a periodic input to the DH system of the date and the time of day.
- *Sensors and Controllers* – Thermostat, Humidistat, Power Switch, Contact Sensor, Light Alarm, Sound Alarm (Figures C.1–C.3).

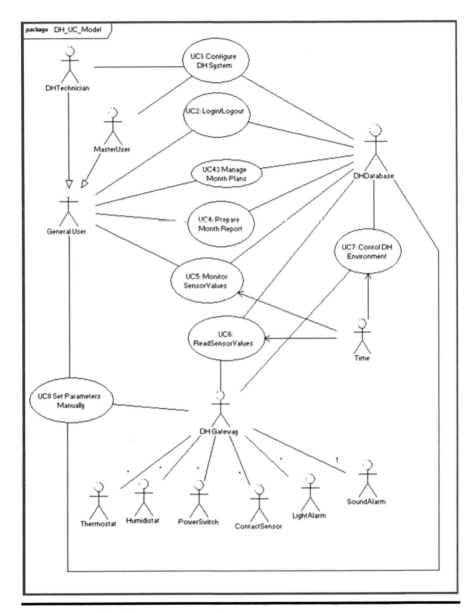

Figure C.1   Digital home use case diagram.

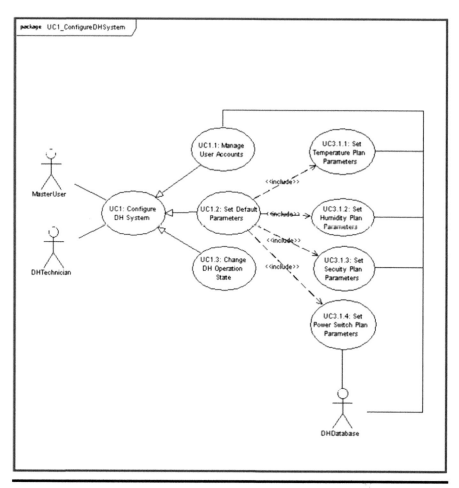

Figure C.2 UC1 configure DH system.

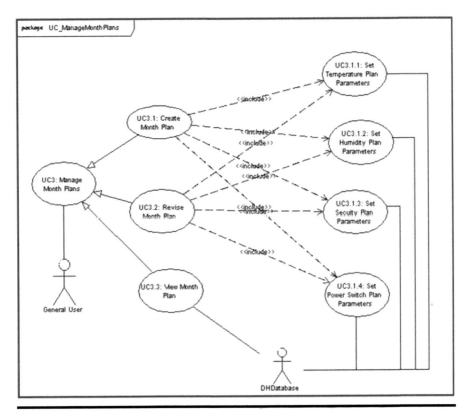

Figure C.3   UC3 manage monthly planner.

UC1: Configure DH System

Use Case ID: UC1

> **Use Case Name**: Configure DH System
>
> **Goal**: Configure the DH system, including adding/deleting and maintaining user accounts, setting/changing default parameters, starting/stopping operation of the DH system.
>
> **Primary Actors**: DH Technician, Master User
>
> **Secondary Actor**: DH Database

Pre

1. DH Technician or Master User is logged into his/her DH account.
2. DH Main Page is displayed on DH Technician or Master User display device.

Post

1. DH Main page is displayed on DH Technician or Master User display device.
2. Requested changes have been saved to DH Database.

Main Success Scenario
UC GUIs: DH Main Page, DH Configuration Page, Account Configuration
Page, Parameters Configuration Page, and System State Configuration Page

Step	Actor Action	Step	System Reaction
1	Select Configuration System option	2	Display DH Configuration Page. Request operation to be performed: a. Add/Delete a user account b. Set/Change Default Values c. Start/Stop DH System Operation d. Exit DH Configuration Page
3A	Select a.	4A	Display Account Configuration Page. *Include* UC1.1 *Go to step 2*
3B	Select b.	4B	Display Default Parameters Configuration.Page. *Include* UC1.2 *Go to step 2*
3C	Select c.	4C	Display System State Configuration Page. *Include* UC1.3 *Go to step 2*
3D	Select d.	4D	Display DH Main Page.

Exceptions (Alternative Scenarios of Failure)
1. At any time, User fails to make an entry.
   a. Time out after five minutes and log out of system.
2. System detects a failure (loss of Internet connection, power loss, hardware failure, etc.).
   a. Display an error message on the User display device.
   b. System restores system data from DH Database.
   c. Log off user.

Use Cases Utilized: UC1.1, UC1.2, UC1.3

Notes and Issues
1. An account is set up for the DH Technician prior to initial system configuration. Such a setup is not part of the normal DH System operation. Master Users and General Users can be added by the DH Technician.

UC1.1: *Manage User Accounts*

Use Case ID: UC1.1

Use Case Name: Manage User Accounts
Goal: Add/delete users to/from the DH System
Primary Actors: DH Technician, Master User
Secondary Actor: DH Database

Pre

1. DH Technician or Master User is logged into his/her DH account.
2. DH Configuration Page is displayed on User display device.

Post

1. DH Configuration Page is displayed on User display device.
2. User account has been added/deleted to/from DH Database.

Main Success Scenario

Step	Actor Action	Step	System Reaction
		1	Request operation to be performed: a. Add a user account b. Delete user account c. Exit Account Configuration Page
2A	Select a.	3A	Request user name and password for new user account.
4A	Enter user name and password	5A	Store account information in DH Database. *Go to step 1.*
2B	Select b.	3B	Request user name for account to be deleted.
4B	Enter user name	4D	Delete account from DH Database. *Go to step 1.*
2C	Select c.	3C	Display DH Configuration Page

UC GUIs: DH Configuration Page, DH Account Configuration

Exceptions (Alternative Scenarios of Failure)
1. User fails to make an entry.
   a. Time out after five minutes and log out of system.
2. System detects a failure (loss of Internet connection, power loss, hardware failure, etc.).
   a. Display an error message on the User display device.
   b. System restores system data from DH Database.
   c. Log off user.

Use Cases Utilized: None.

Notes and Issues
It not possible to prevent all errors concerned with creation of user accounts – e.g., incorrect user name or unintended password.

UC1.2: Set Default Parameters

Use Case ID: UC1.2
  Use Case Name: Set Default Parameters
  Goal: Set/change default parameters for DH sensors and controllers
  Primary Actors: DH Technician, Master User
  Secondary Actor: DH Database

Pre
1. DH Technician or Master User is logged into his/her DH account.
2. DH Default Parameter Configuration Page is displayed on User display device.

Post
1. All default parameters have been stored in DH Database.
2. DH Configuration Page is displayed on User display device.

Main Success Scenario

Step	Actor Action	Step	System Reaction
		1	Display "Enter the default set point temperature for all Thermostats".
2	Enter set point temperature for each Thermostat.	3	Store the temperature set points in the DH database.

(Continued)

Step	Actor Action	Step	System Reaction
		4	Display "Enter the default set point humidity for all Humidistats".
5	Enter set point humidity for each Humidistat.	6	Store the humidity set points in the DH database.
		7	Display "Enter the default setting for all of the contact sensors (ON/OFF)".
8	Enter On or OFF	9	Store the contact sensor setting in the DH database.
		10	Display "Select room for setting power switch defaults".
11	Select room	12	Display "Select power switch(es)" and "For power switches not selected, they will be set to OFF".
13	Select power switch(es)	14	Store the power switch settings, for the selected room, in the DH database.
		15	Display "For rooms not selected, all power switches will be set to OFF. Other rooms? Yes/No".
16	Enter "Yes"	17	Go to Step 10.
18	Enter "No"	19	Display DH Configuration Page.

UC GUIs: DH Configuration Page, DH Default Parameter Configuration Page

Exceptions (Alternative Scenarios of Failure)

1. At any time, User fails to make an entry.
   a. Time out after five minutes and log out of system.
2. System detects a failure (loss of Internet connection, power loss, hardware failure, etc.).
   a. Display an error message on the User display device.
   b. System restores system data from DH Database.
   c. Log off user.

Use Cases Utilized: None

UC1.3: Change DH System Operation State

Use Case ID: UC1.3

Use Case Name: Change DH System Operation State
Goal: Start/Stop DH System Operation
Primary Actors: DH Technician, Master User
Secondary Actor: DH Database

Pre

1. DH Technician or Master User is logged into his/her DH account.
2. DH System State Configuration Page is displayed on DH Technician or Master User display device.

Post

1. DH Configuration Page is displayed on DH Technician or Master User display device.
2. DH System state has been stored in DH Database.

Main Success Scenario

Step	Actor Action	Step	System Reaction
		1	Retrieve state of DH System operation from DH Database.
		2	Display "ON or OFF" for state of DH System operation. Display "Change the state of DH operation?" YES/NO.
3A	Select YES.	4A	Store new state of DH system in the DH Database.
3B	Select NO.	4B	Display DH Configuration Page.

UC GUIs: DH Configuration Page, DH System State Configuration Page

Exceptions (Alternative Scenarios of Failure)

1. At any time, User fails to make an entry.
   a. Time out after five minutes and log out of system.
2. System detects a failure (loss of Internet connection, power loss, hardware failure, etc.).
   a. Display an error message on the User display device.
   b. System restores system data from DH Database.
   c. Log off user.

Use Cases Utilized: None

UC2: Login/Logout

Use Case ID: UC2

    Use Case Name: Login/Logout

    Goal: Login/logout of the DH system; change the password.

    Primary Actor: General User

    Secondary Actor: DH Database

Pre

    1. Actor has a user account.

    2. DH Main Page is displayed.

Post

    1. Login/Logout and/or password change is successful

    2. DH Main Page is displayed.

Main Success Scenario

Step	Actor Action	Step	System Reaction
1A	Select Login option.	2A	Display Login Page. Request to enter username and password.
3A	Enter user name and password.	4A	Check user name and password in DH Database.
		5A	If incorrect, print error message and *go to step 2A*. If correct, print login confirmation message and *go to step 6A*.
		6A	Display DH Main Page
7A	Select Change password option.	8A	Request to enter the current and new password.
9A	Enter the current and new password.	10A	Store new password in the DH Database. Print password change confirmation message.
1B	Select Logout option.	2B	Print Logout confirmation message.
		3A	Display DH Main Page.

UC GUIs: DH Main Page, Login/Logout Page

Exceptions (Alternative Scenarios of Failure)
1. User fails to make an entry.
   a. Time out after five minutes and log out of system.
2. A user, who is already logged in, tries to log in.
   a. Display an error message. *Go to 1A.*
3. A user, who is logged out, tries to log out.
   a. Display an error message. *Go to 1A.*
4. User enters non-existent username/password combination five consecutive times.
   a. Display error message and log out of system
   b. Time out users for 10 minutes.
5. System detects a failure (loss of Internet connection, power loss, hardware failure, etc.).
   a. Display an error message on the User display device.
   b. System restores system data from DH Database.
   c. Log off user.

Use Cases Utilized: None.

UC3: Manage Monthly Planner

Use Case ID: UC3

    Use Case Name: Manage Monthly Planner

    Goal: Create, view, or modify monthly plans through which control the environment of a Digital Home.

    Primary Actor: General User

    Secondary Actor: DH Database

Pre
1. User is logged into his/her DH account
2. DH Main Page is displayed.

Post
1. Month Plan operation has been completed
2. DH Main Page is displayed on the user display device.

Main Success Scenario

Step	Actor Action	Step	System Reaction
1	Select Manage Month Plan option.	2	Display Month Plan Page and Request to select an operation to be performed: a. Create new month plan. b. Modify existing month plan. c. View existing month plan. d. Exit Month Plan Page.
5A	Select a.	6A	*Include* UC4.1, then *Go to step 2.*
5B	Select b.	6B	*Include* UC 4.2, then *Go to step 2.*
5C	Select c.	6C	*Include* UC 4.3, then Go to *step 2.*
5D	Select d.	6D	Display DH Main Page.

UC GUIs: DH Main Page, Month Plan Page

Exceptions (Alternative Scenarios of Failure)
1. User fails to make an entry.
   a. Time out after five minutes and log out of system.
2. System detects a failure (loss of Internet connection, power loss, hardware failure, etc.).
   a. Display an error message on the User display device.
   b. System restores system data from DH Database.
   c. Log off user.

Use Cases Utilized: UC3.1, UC3.2, and UC3.3

Notes and Issues
System in the digital home is a self-sustaining wireless system that continues to function during the loss of Internet connection using the latest update of sensors and controllers values until the connection is re-established.

UC3.1: Create Month Plan

Use Case ID: UC3.1
   Use Case Name: Create Month Plan
   Goal: Create the month plan for setting and controlling environmental parameters (temperature, humidity, contact sensors and power switches).
   Primary Actor: General User
   Secondary Actor: DH Database

Pre
1. Default parameters for the system have been set.
2. User has selected Create Month Plan option

Post
1. Month Plan has been created and stored in DH Database.

Main Success Scenario

Step	Actor Action	Step	System Reaction
		1	Request Month and Year.
2	Enter Month and Year.	3	Default environmental parameters are entered for all days, and a message about this is displayed.
		4	Display the Thermostat Month Plan Page.
5	*Include* UC4.1.1 (Set Temperature Plan Parameters)		
		6	Display the Humidity Month Plan Page.
7	*Include* UC4.1.2 (Set Humidity Plan Parameters)		
		8	Display the Security Month Plan Page.
9	*Include* UC4.1.3 (Set Security Plan Parameters)		
		10	Display the Power Switch Month Plan Page.
11	*Include* UC4.1.4 (Set Power Switch Plan Parameters)		

UC GUIs: Thermostat Month Plan Page, Humidistat Month Plan Page, Security Month Plan s Page, and Power Switch Month Plan Page

Exceptions (Alternative Scenarios of Failure)
1. User fails to make an entry.
   a. Time out after five minutes and log out of system.
2. A plan already exists for the month and year entered.
   a. Print a message about the existence of the plan and Display DH Main Page.

3. System detects a failure (loss of Internet connection, power loss, hardware failure, etc.).
   a. Display an error message on the User display device.
   b. System restores system data from DH Database.
   c. Log off user.

Use Cases Utilized: UC3.1.1, UC3.1.2, UC3.1.3, and UC3.1.4

UC3.1.1: Set Temperature Plan Parameters

Use Case ID: UC3.1.1

    Use Case Name: Set Temperature Plan Parameters

    Goal: Set the thermostat set points for time periods for each thermostat, for a specified month/year and day(s) for different home spaces.

    Primary Actor: General User

    Secondary Actor: DH Database

Pre

1. User has specified month and year for temperature settings.
2. Thermostat Month Plan Page is displayed.

Post

1. Set points are set for the specified thermostats periods, and days during the specified month and year.

Main Success Scenario

Step	Actor Action	Step	System Reaction
		1	Request which Thermostats are to be set.
2	Enter Thermostat designations.	3	Request days to be scheduled.
4	Enter days.	5	Request to enter (up to 4) time periods (start/end) for thermostats.
5	Enter time periods (start/end).	6	Request to set temperature values for thermostats in Fahrenheit or Celsius.

(Continued)

Step	Actor Action	Step	System Reaction
7A	Enter Fahrenheit.	8A	Request to set temperature values in range between 60°F and 80°F.
7B	Enter Celsius.	8B	Request to set temperature values in range between 16°C and 27°C.
9	Enter values for the selected days and periods.	10	Request to proceed to different Humidistats and/or days. YES/NO?
11A	Select YES.	12A	*Go to Step 1.*
11B	Select NO.	12B	Save day/period/set point values to the Thermostat Month Plan in DH Database.
		13	Display confirmation message.

UC GUIs: DH Main Page, Thermostat Month Plan Page

Exceptions (Alternative Scenarios of Failure)
1. User fails to make an entry.
   a. Time out after five minutes and log out of system.
2. System detects a failure (loss of Internet connection, power loss, hardware failure, etc.).
   a. Display an error message on the User display device.
   b. System restores system data from DH Database.
   c. Log off user.

Use Cases Utilized: None.

Notes and Issues
Days with no values entered will be set to the default values or previously set values (if any).

UC3.1.2: Set Humidistat Plan Parameters

Use Case ID: UC3.1.2

   Use Case Name: Set Humidistat Plan Parameters

   Goal: Set the humidity set points for the time periods for humidistats, for specified month and year and day(s) for different home spaces.

Primary Actor: General User
Secondary Actor: DH Database

Pre
1. User has specified month and year for humidity settings.
2. Humidistat Month Plan Page is displayed.

Post
1. Set points are set for the specified Humidistats, periods, and days during the specified month and year.

Main Success Scenario

Step	Actor Action	Step	System Reaction
		1	Request which Humidistats are to be set.
2	Enter Humidistat designations.	3	Request days to be scheduled.
4	Enter days.	5	Request to enter (up to 4) time periods (start/end) for Humidistats.
5	Enter time periods (start/end).	6	Request to set humidity values for specified days and periods.
7	Enter values for the selected days and periods.	8	Request to proceed to different Humidistats and/or days. YES/NO?
9A	Select YES.	10A	*Go to Step 1.*
9B	Select NO.	10B	Save day/period/set point values to the Humidistat Month Plan in DH Database.
		11	Display confirmation message.

UC GUIs: DH Main Page, Humidistat Month Plan Page

Exceptions (Alternative Scenarios of Failure)
1. User fails to make an entry.
   a. Time out after five minutes and log out of system.

2. System detects a failure (loss of Internet connection, power loss, hardware failure, etc.).
   a. Display an error message on the User display device.
   b. System restores system data from DH Database.
   c. Log off user.

Use Cases Utilized: None.

Notes and Issues

Days with no values entered will be set to the default values or previously set values (if any).

UC3.1.3: Set Security Plan Parameters

Use Case ID: UC3.1.3

Use Case Name: Set Security Plan Parameters

Goal: Set Contact Sensors to ARMED/UNARMED for specified time periods for a specified month and day(s). **Primary Actor**: General User

Secondary Actor: DH Database

Pre

1. User has specified month and year for contact sensor settings.
2. Security Month Plan Page is displayed.

Post

1. ARMED or UNARMED set for designated contact sensors, for specified periods and days during the specified month and year.

Main Success Scenario

Step	Actor Action	Step	System Reaction
		1	Request which contact sensors are to be set.
2	Enter Contact Sensor designations.	3	Request days to be scheduled.
4	Enter days.	5	Request to enter (up to 4) time periods (start/end) for contact sensors.

(Continued)

Step	Actor Action	Step	System Reaction
5	Enter time periods (start/end).	6	Request to set contact sensor values for specified days and periods.
7	Enter values for the selected days and periods.	8	Request to proceed to different contact sensors and/or days. YES/NO?
9A	Select YES	10A	*Go to Step 1.*
9B	Select NO	10B	Save day/period values to the Security Month Plan in DH Database.
		11	Display confirmation message.

UC GUIs: DH Main Page, Security System Parameters Plan Page

Exceptions (Alternative Scenarios of Failure)
1. User fails to make an entry.
   a. Time out after five minutes and log out of system.
2. System detects a failure (loss of Internet connection, power loss, hardware failure, etc.).
   a. Display an error message on the User display device.
   b. System restores system data from DH Database.
   c. Log off user.

Use Cases Utilized: None.

Notes and Issues
Days with no values entered will be set to the default values or previously set values (if any).

UC3.1.4: Set Power Switch Plan Parameters

Use Case ID: UC3.1.4

 Use Case Name: Set Power Switch Plan Parameters

 Goal: Set Power Switches to ON/OFF for specified time periods for a specified month and day(s).

 Level: User-goal

 Primary Actor: General User

 Secondary Actor: DH Database

Pre

1. User has specified month and year for power switch settings.
2. Power Switch Month Plan Page is displayed.

Post

1. ON or OFF set for designated power switches, for specified periods and days during the specified month and year.

Main Success Scenario

Step	Actor Action	Step	System Reaction
		1	Request which power switches are to be set.
2	Enter Power Switch designations.	3	Request days to be scheduled.
4	Enter days.	5	Request to enter (up to 4) time periods (start/end) for power switches.
5	Enter time periods (start/end).	6	Request to set power switch values for specified days and periods.
7	Enter values for the selected days and periods.	8	Request to proceed to different contact sensors and/or days. YES/NO?
9A	Select YES.	10A	*Go to Step 1.*
9B	Select NO.	10B	Save day/period values to the Power Switch Month Plan in DH Database.
		11	Display confirmation message.

UC GUIs: DH Main Page, Power Switch Plan Page

Exceptions (Alternative Scenarios of Failure)

1. User fails to make an entry.
   a. Time out after five minutes and log out of system.
2. System detects a failure (loss of Internet connection, power loss, hardware failure, etc.).
   a. Display an error message on the User display device.
   b. System restores system data from DH Database.
   c. Log off user.

Use Cases Utilized: None.

Notes and Issues
Days with no values entered will be set to the default values or previously set values (if any).

UC3.2: Modify Month Plan

Use Case ID: UC3.2

    Use Case Name: Modify Month Plan

    Goal: Modify an existing month plan for setting and controlling environmental parameters (temperature, humidity, contact sensors, and power switches).

    Primary Actor: General User

    Secondary Actor: DH Database

Pre
1. User has selected Modify Month Plan option

Post
1. Month plan has been modified and stored in DH Database

Main Success Scenario

Step	Actor Action	Step	System Reaction
		1	Request Month and Year.
2	Enter Month and Year.	3	Retrieve month plan from DH Database.
		4	Display the Thermostat Month Plan Page.
5	Include UC4.1.1 (Set Temperature Plan Parameters)		
		6	Display the Humidity Month Plan Page.
7	Include UC4.1.2 (Set Humidity Plan Parameters)		
		8	Display the Security Month Plan Page.
9	Include UC4.1.3 (Set Security Plan Parameters)		
		10	Display the Power Switch Month Plan Page.
11	Include UC4.1.4 (Set Power Switch Plan Parameters)		

UC GUIs: Thermostat Month Plan Page, Humidity Month Plan Page, Security Month Plan Page, Power Switch Month Plan Page

Exceptions (Alternative Scenarios of Failure)
1. User fails to make an entry.
    a. Time out after five minutes and log out of system.
2. A plan does not exist for the month and year entered.
    a. Print a message about the non-existence of the plan and Display DH Main Page.
3. System detects a failure (loss of Internet connection, power loss, hardware failure, etc.).
    a. Display an error message on the User display device.
    b. System restores system data from DH Database.
    c. Log off user.

Use Cases Utilized: UC3.1.1, UC3.1.2, UC3.1.3, and UC3.1.4

UC3.3: *View Month Plan*

Use Case ID: UC3.3

   Use Case Name: View Month Plan

   Goal: View existing months plan for setting and controlling environmental parameters (temperature, humidity, contact sensors and power switches).

   Primary Actors: Master User, General User

   Secondary Actor: DH Database

Pre

   1. User has selected View Month Plan option

Post

   2. None.

Main Success Scenario

Step	Actor Action	Step	System Reaction
		1	Request Month and Year.
2	Enter Month and Year.	3	Retrieve month plan from DH Database.
		4	Request to select an option: a. view Thermostat Month Plan b. view Humidistat Month Plan c. view Security System Month Plan d. view Power Switch Month Plan e. exit

(Continued)

Step	Actor Action	Step	System Reaction
5A	Select a.	6A	Display Thermostat Month Plan.
5B	Select b.	6B	Display Humidistat Month Plan.
5C	Select c.	6C	Display Security System Month Plan.
5D	Select d.	6D	Display Power Switch Month Plan.
5E	Select e.	6E	Return (to UC4).

UC GUIs: Thermostat Month Plan Page, Humidity Month Plan Page, Security Month Plan Page, Power Switch Month Plan Page

Exceptions (Alternative Scenarios of Failure)

1. User fails to make an entry.
   a. Time out after five minutes and log out of system.
2. A plan does not exist for the month and year entered.
   a. Print a message about the non-existence of the plan and Display DH Main Page.
3. System detects a failure (loss of Internet connection, power loss, hardware failure, etc.).
   a. Display an error message on the User display device.
   b. System restores system data from DH Database.
   c. Log off user.

Use Cases Utilized: None.

UC4: Prepare Month Report

Use Case ID: UC4

Use Case Name: Prepare Month Report

Goal: Prepare a DH monthly report that includes data about temperature and humidity values, power switches states, security breaches, and error messages.

Primary Actor: General User

Secondary Actor: DH Database

Pre

1. User is logged into his/her DH account.
2. DH Main Page is displayed on the user display device.

Post

1. A monthly report has been prepared.
2. DH Main Page is displayed on the user display device

Main Success Scenario

Step	Actor Action	Step	System Reaction
1	Select Monthly Report option.	2	Request to enter month and year in the past two years.
3	Enter valid month and year.	4	Retrieve information about requested month and year from DH Database.
		5	Compute the following values: for each thermostat, average temperature for each day of the month, times and values of the maximum and minimum temperatures for each day.
		6	Compute the following values: for each humidistat, average humidity for each day of the month, times and values of the maximum and minimum temperatures for each day.
		7	Determine information about any security breaches (contact sensors ID and date/time of occurrence).
		8	Determines information about any periods (date/time and time interval) when the DH System was not in operation.
		9	Retrieve information associated with error messages.
		10	Use information from steps 4–9 to generate the report.
		11	Display the report.
		12	Requests "Exit"
13	Enter "Exit"		
		14	Display DH Main Page.

UC GUIs: DH Main Page, Monthly Report Page

Exceptions (Alternative Scenarios of Failure)
1. User fails to make an entry.
   a. Time out after five minutes and log out of system.

2. Data not available for Month/Year entered.
   a. Display an error message on the User display device.
   b. Display DH Main Page
3. System detects a failure (loss of Internet connection, power loss, hardware failure, etc.).
   a. Display an error message on the User display device.
   b. System restores system data from DH Database.
   c. Log off user.

Use Cases Utilized: None.

UC5: Monitor Sensor Values

Use Case ID: UC5

Use Case Name: Monitor Sensor Values

Goal: Get and display the current values of the temperature, humidity, states of contact sensors, light and sound alarms, and states of power switches.

Primary Actor: Master User, General User

Secondary Actor: DH Database, Time

Pre
1. User is logged into DH account.
2. DH Main Page is displayed on the user display device.

Post
1. Current sensor values are displays on DH Sensor Values Page

Main Success Scenario

Step	Actor Action	Step	System Reaction
1	Request display of sensor values.	2	Get values, from DH Database, of temperature, humidity, states of power switches, states of contact sensors, light and sound alarms.
		3	Display values and states on DH Sensor Values Page.
		4	Delay ΔT.
5A	No exit page request.	6A	*Repeat steps 2–4.*
5B	Exit page request.	6B	Display DH Main Page.

UC GUIs: DH Main Page, Sensor Values Page

Exceptions (Alternative Scenarios of Failure)
1. User fails to make an entry.
   a. Time out after five minutes and log out of system.
2. System detects a failure (loss of Internet connection, power loss, hardware failure, etc.).
   a. Display an error message on the User display device.
   b. System restores system data from DH Database.
   c. Log off user.

Use Cases Utilized: None.

Notes and Issues
1. System in the digital home is a self-sustaining wireless system that continues to function during the loss of Internet connection using the latest update of sensors and controllers values until the connection is re-established.
2. The ΔT delay time would be chosen appropriate for the database read/update times.
3. Time is a virtual actor that provides the current date and time.

UC6: Read Sensor Values

Use Case ID: UC6

   Use Case Name: Read Sensor Values

   Goal: Continuously read the current values of the temperature, humidity, state of contact sensors, states of power switches and store the values and states in DH Database.

   Primary Actor: Time

   Secondary Actors: DH Gateway, DH Database

Pre
1. Operation of the DH System has been initiated.

Post
1. Values from Sensors and Devices have been saved in DH Database.
2. Operation of the DH System has been terminated.

Main Success Scenario

Step	Actor Action	Step	System Reaction
1	Date and time provided.	2	Read values, through DH Gateway, of temperature, humidity, states of power switches, state of contact sensors, and security sound and light alarm states.
		3	Write sensor values, alarm states, and date and time in DH Database.
		4	Delay ΔT.
		5	*Go to step 1.*

UC GUIs: None.

Exceptions (Alternative Scenarios of Failure)
1. System detects a failure (loss of Internet connection, power loss, hardware failure, etc.).
   a. Display an error message on the User display device.
   b. System restores system data from DH Database.
   c. Log off user.

Use Cases Utilized: None.

Notes and Issues
1. The ΔT delay time would be chosen appropriate for the sensor update/ read times and database write times.
2. Time is a virtual actor that provides the current date and time.

UC7: Control DH Environment

Use Case ID: UC7

    Use Case Name: Control DH Environment

Goals
1. Using the current month plan or the default values, set the values for the temperature, humidity, contact sensors states, and power switches states.
2. Activate the security alarms (one sound alarm and one light subsystem with multiple lights) if a contact sensor is breached.

Primary Actor: Time

    Secondary Actors: DH Gateway, DH Database

Pre

1. Default settings or Month plan for current month (if any) have been created and saved in DH Database.
2. Operation of the system is initiated (system state is in ON state).

Post

1. Operation of DH System is terminated (DH Database contains an OFF operation state)

Main Success Scenario

Step	Actor Action	Step	System Reaction
1	Date and time provided.	2	Get, from DH Database, month plan for current date and read controller values for current time. If no such plan exists, get default values for current data and time.
		3	Send controller values to controllers, through the DH Gateway.
		4	Check if a contact sensor has been breached and if so, activate alarms.
		5	Delay ΔT.
		6	*Go to step 1.*

UC GUIs: None.

Exceptions (Alternative Scenarios of Failure)

1. System detects a failure (loss of Internet connection, power loss, hardware failure, etc.).
   a. Display an error message on the User display device.
   b. System restores system data from DH Database.
   c. Log off user.

Notes and Issues

1. The ΔT delay time would be chosen appropriate for the sensor update/ read times and database write times.
2. Time is a virtual actor that provides the current date and time.

UC8: Set Parameters Manually

Use Case ID: UC8

Use Case Name: Set Parameters Manually

Goal: Set Thermostat, Humidistat parameters and Appliances states manually, without accessing User account directly.

Primary Actor: General User

Secondary Actors: Thermostat, Humidistat, Power Switch, Security System (Contact Sensors), DH Gateway, DH Database.

Pre

1. Current parameters or states are displayed at a device location.

Post

2. Manual settings have been updated at a device location.
3. _Manual settings have overridden planned/default settings in DH Database for the current period.

Main Success Scenario

Step	Actor Action	Step	System Reaction
1A	View and set temperature value manually at a Thermostat location in specified range.	2A	Update temperature value at a Thermostat location.
		3A	Override temperature value for the Thermostat in DH Database for the current period.
1B	View and set Humidistat level manually at a Humidistat location in specified range.	2B	Update humidity level at a Humidistat location.

(Continued)

Step	Actor Action	Step	System Reaction
		3B	Override humidity level for the Humidistat in DH Database for the current period.
1C	View and set the state of Power Switch manually at a device location.	2C	Update Power Switch state at a device location.
		3C	Override Power Switch state for the device in DH Database for the current period.
1D	View and set the state of Contact Sensor(s) at a device location.	2D	Update the state of Contact Sensor(s) at a device location.
		3D	Retrieve information from the DH Database about the starting time and state of contact sensors at a device location for the next time period. Display a descriptive message.
		4D	Override Contact Sensors state for the device in DH Database for the current period.

UC GUIs: None.

Exceptions (Alternative Scenarios of Failure)
1. System detects a failure (loss of Internet connection, power loss, hardware failure, etc.).
   a. Display an error message on any User display devices.
   b. System restores system data from DH Database.
   c. Log off any online users.

Use Cases Utilized: None.

Notes and Issues
System in the digital home is a self-sustaining wireless system that continues to function during the loss of Internet connection using the latest update of sensors and controllers values from the DH Database or manually entered values until the connection is re-established.

Index

Note: **Bold** page numbers refer to tables; *italic* page numbers refer to figures.

abstract factory **187**
accreditation board for engineering and
    technology (ABET) 5, 239,
    250–251
accreditation criteria 5
adapter **187**
agile processes 29–30
alpha testing 79
analyst, software 8–9
appraisal cost 57
architect, software 9
architecture, software design
    allocation view 164
    client-server style *168,* 168–169
    code view 166
    component and connector (C&C) view 164
    conceptual view 165
    definitions 160
    deployment view 163
    development view 164
    documenting decisions 161
    entities referred 160
    execution view 166
    importance 159–160
    Krutchen's view 163–164
    layered style 169, *169*
    logical view 163
    4+1's logical view 165–166
    module view 164, 166
    physical view 163
    pipe and filter style 167–168
    process view 163
    quality attributes 160–161
    SEI views 164, *165*
    service-oriented style 170, *170*
    Siemens four-views approach 165–166
    and stakeholders 162

styles of 167–170
use cases view 164
views 161–166
association for computing machinery (ACM)
    5, 242

backlog
    buffer 262
    epics 261
    estimation techniques 264–265
    items 261
    refinement/grooming 263
    sprint backlog 261–262
    stories 261
    tasks 261
    themes 261
beta testing 79
black box techniques, unit testing 75
Boehm, B. 29, 31
Booch, G. 181
bridge **187**
British computer society (BCS) 239
Brooks, F. P. 4, 5, 124
Brooks, J. 238
builder **187**
burnup chart 271
business risks 106

Canadian engineering accreditation board
    (CEAB) 239
capability maturity model (CMM) 34, *34*
capability maturity model integration (CMMI)
    6, 34, *34*
case study
    appropriate set of views selection, DH 192
    architectural style selection, DH 192–193
    architecture development, DH 193–194

331

case study (*cont.*)
  assessing customer needs 20–21
  code review 84–85
  configuration identification 115, 119–120
  configuration management 115–116, 120
  cost of quality experiences 80–81
  design concepts and principles 191
  design patterns application 197–198
  design quality measurement 194–196
  development strategy, DH 46–47
  documenting, software design 198–199
  effective team formation 45
  eliciting requirements, DH 144–145
  ethical issues, DH 251–253
  evaluating a software process 45–46
  goal/question/metric, DH 82–83
  integration test planning 86–87
  legal issues, DH 254
  maintenance costs estimation, DH 233–234
  maintenance process determination,
      DH 233
  potential maintenance problems
      identification, DH 231–232
  quality attribute specification 143–144
  quality measurement strategy, DH 82
  quality planning 116, 120–121
  re-engineering effort, DH 234–235
  requirements inspection 83–84
  risk assessment 109, 118–119
  security attacks 79–80
  security requirements 148–149
  software quality assurance planning 81–82
  standards determination, DH 253–254
  task plan and schedule 105, 117–118
  team member hire, DH 250
  team problems 17–20
  unit test planning 85–86
  use case modeling, DH 146–147
  use case test planning *77–78*, 88–89
  writing software requirements 147–148
certified software development professional
    (CSDP) 6, 241
chain of responsibility **188**
changeability 5
Chidamber and Kemerer metrics 182–184
Chidamber, S. R. 182
client groups 3
client-server architectural style *168,* 168–169
COCOMO *see* constructive cost model
    (COCOMO)
code restructuring 230–231
code review 72, 84–85, 212, **213–214**

coding standards 37, 211–212
collective ownership, extreme programming 37
competitor 2
complexity 5
composite **187**
computer emergency response team (CERT) 214
concept of operations (ConOps) document 16–17
configuration management *see* software
    configuration management (SCM)
conformity 5
construction principles
  buffer overflow problem 219–220
  build/integration plan 201–205, 216
  code review 212, **213–214**
  coding standards 211–212
  computer emergency response team
      (CERT) 214
  design diagram, DH **205**
  development process script **202–204,**
      218–219
  fundamentals 206–208
  GCD pseudocode 210–211
  java code conventions 211–212
  post-condition 209, 217
  pre-condition 208, 217
  pseudocode 209–211
  secure coding **215**
  specification 208, 216, 217
  test-driven development (TDD) 209
  test plan/report *210,* 218
  unified process 35
  unit testing 209
constructive cost model (COCOMO) 100–101
content coupling modules 158
continuous integration, extreme programming 37
controlled centralized decision 27
controlled decentralized decision 28
copyrights 248
cost of quality 57–58, 80–81
critical-path method (CPM) 102–103
curriculum, software engineering programs
    criteria 240
customer need assessment 17, 20–21
customer satisfaction, quality metrics 65

daily scrum 38
data coupled modules 158
decorator **187**
defect tracking 60–61
defects, software engineering 4
Delphi software estimation method 98–99
DeMarco, T. 182

democratic decentralized decision 27
design, concepts and principles; *see also* object-
        oriented design
    abstraction 155
    assuring design quality 181–182
    cohesion 157
    component and module 153
    coupling 158
    definition 153
    design documentation 189–190
    design verification 181–184
    encapsulation 155, *156*
    information hiding 157
    modularity 153, *154*
    object-oriented design 171–181
    phases, DH **152**
    quality assurance 181–184 (*see also* software
            quality assurance (SQA))
    security 158–159
    software architecture (*see* architecture,
            software design)
design restructuring 230
development team 8–10
DigitalHome (DH) system
    appropriate set of views selection 192
    architectural style selection 192–193
    architecture development 193–194
    business rules 297
    change history 287–288
    communication center 14
    concept of operations 16–17
    configuration identification 115, 119–120
    configuration management 115–116, 120
    cost of quality experiences 80–81
    customer need assessment 17
    customer need statement 13, 283–286
    design concepts and principles 191
    design patterns application 197–198
    design quality measurement 194–196
    development constraints 291
    development process script **40–42**
    development strategy 42–44, 46–47
    discussion, debate and decisions 2–3
    documenting, software design 198–199, 298
    eliciting requirements 132, 144–145
    functional requirements 293–296
    general requirements 293
    GQM technique evaluation 82–83
    high-level requirements definition 13–14
    humidistat requirements 15, 294–295
    maintenance costs estimation 233–234
    maintenance process determination 233

    maintenance requirements 297
    management requirements 295
    master control device 14
    needs assessment 12–13
    non-functional requirements 296–298
    operational environment 292–293
    performance requirements 296
    planner, parameters set 15–16, 295–296
    product description and scope 290
    programmable power switch 15
    project team 10–13
    prototype features 14–16
    quality attribute specification 143–144
    quality measurement strategy 82
    quality planning 116, 120–121
    re-engineering effort 234–235
    reliability 296
    risk assessment 109, 118–119
    safety requirements 297
    security engineering 79–80
    security requirements 148–149, 295, 297
    security system 14
    sensors 14
    smart features 1–2
    software quality assurance plan 81–82
    specification, requirements 148–149,
            289–298
    task plan and schedule 105, 117–118
    team project information 289–290
    thermostat requirements 14–15, 294
    trend presentation 2
    use case modeling 132–136, 146–147
    users description 290–291
    writing software requirements 147–148
documentation
    architecture document uses 161
    concept of operations (ConOps) 16–17
    description standards 189–190
    design 180, 189–190
    multiple viewpoint 189
    object-based architectural design 179
dynamic evaluation 66

earned value tracking 103–104
ease of use, quality metrics 65
elaboration, unified process 35
Institute of Electrical and Electronics Engineers
        Computer Society (IEEE-CS) 5, 242
eliciting requirements
    apprenticing 130
    interviews 130
    observation 130–131

eliciting requirements (*cont.*)
    prototyping 131
    storyboarding 131
    use case 131
endurance test 77
engineering accreditation commission (EAC) 239
ethics; *see also* professionalism
    IEEE Code 244
    importance 243
    SE code 244–245
European network for accreditation of
    engineering education 239
evaluation 39, 45–46, 66, 82–83
expert judgment method 98
extreme programming (XP) 35–37

façade **187**
faculty, software engineering programs criteria
    240
Fagan inspection process
    follow-up activity 71
    formal meeting 71
    individual inspection phase 70
    inspection analysis 71–72
    overview phase 70
    preparation 70
    rework phase 71
    work product 70
Fagan, M. 70
failure cost 57
flyweight **187**
formal specification languages **140**
format
    development strategy form 47
    meeting preparation form 17, 20–21
formulating plans 2–3
forward engineering 231
functional structure project management **95**
function point 98

Gantt chart 102–103, *103*
GQM technique, quality assurance 61–63
graphical notation **140**

high-level requirements definition (HLRD)
    13–14

IEEE Code Ethics 244
IEEE computer society 242
IEEE standard, software quality assurance
    process 59–60
IEEE Std 830-1998 246

IEEE Std 1012™-2012 247
IEEE Std 1016™-2009 246
IEEE Std 12207-2008 246
IEEE Std 14764-2006 247
impact analysis 228
implementation languages 157
inception, unified process 35
incremental-model *33*, 33–34
increment releases, unified process 35
inspection, quality assurance
    examination factors 69
    Fagan inspection process 70–72
    participants, inspection meeting 70
    requirements inspection 72, 83–84
    vs. technical review 69–70
integration tests 75–76, 79, 86–87
intellectual property (IP) 248
interpreter **188**
invisibility 5
ISO/IEC/IEEE 24765 246
iterator **188**

Japan accreditation board for engineering
    (JABEE) 239

Kaizen 274
Kemerer, C. F. 182
knowledge areas (KA), SWEBOK **6–8**
Kruchten, P. 163
Krutchen's view architecture 163–164

launch process, software project
    script particulars **24**
    team member meeting 23–24
Lawford, M. 249
layered architectural style 169, *169*
legacy system 229
legal issues
    copyrights 248
    intellectual property 248
    liability focus 249
    negligence harms 247, 249
    patents 248
    trade secret 248–249
liability legal issues 249
lines of code (LOC) 98

maintenance
    adaptive maintenance situation 223
    analyzability 226
    backlog management index 225
    categories 222–223

changeability 226
code restructuring 230–231
corrective maintenance 222–223
costs 225
definition 222
design restructuring 230
document restructuring 230
forward engineering 231
impact analysis 228
inventory analysis 230
legacy system 229
measurement 225–226
migration 227
modification implementation 227
objective of 222
perfective maintenance situation 223
preventive maintenance situation 223
problem and modification analysis 226
process activities 226–228, *227*
process implementation 226
re-engineering process 229
restructuring 230–231
retirement 227
reverse engineering 231
review/acceptance 227
and software evolution 222
stability 226
testability 226
vs. hardware maintenance 222
market research 2
mathematical notation **140**
McCabe, T. J. 183
mediator **188**
memento **188**
metaphor, extreme programming 37
metrics, quality
during development 63–65
post release 65
milestone setting 26
models and methods
capability maturity model (CMM) 34, *34*
incremental-model *33*, 33–34
spiral model *31*, 31–32
V-model *32*, 32–33
waterfall model *30*, 30–31

natural language **140**
negligence legal issues 247, 249

object constraint language **140**
object-oriented analysis diagrams, SRS **140**
object-oriented design 171

aggregation relationships 172
association relationships 172
behavioral patterns **188**
Chidamber and Kemerer metrics 182–184
class diagram 172–174
composition relationships 172, *173*
coupling between object classes (CBO) 183
creational design patterns **187**
cyclomatic complexity *183*
define semantics **179**
design external interfaces **179**
design patterns 184–188
develop operational scenarios **179**
documentation **179**
establish dependencies **179**
evaluate design **180**
exit criteria **180**
identify packages and classes **178**
inheritance relationship 174, *174*
lack of cohesion of methods (LCOM) 184
measure quality 182–184
modeling language 171
modules 171
object class 171
package 171
process script **178–180**
response for a class (RFC) 183
revise design **180**
sequence diagram 174–176, *175*
software reuse 184–188
state machine diagrams 176–177
weighted methods per class (WMC) 182–183
observer **188**
organizational structure project management
95–96

pair programming, extreme programming 37
Parnas, D. L. 249
patents 248
perfective maintenance situation 223
performance test 77
phases, software development process
acceptance test **42**
analysis **40**
architectural design **41**
construction increment **41**
exit criteria **42**
inception **40**
launch **40**
planning **40**
postmortem **42**
system test **41**

pipe and filter architectural style 167–168
plan-driven processes 29–30
planning game, extreme programming 36
planning manager 10
presentation
    guidance proposal 2
    market trend 2
prevention cost 57
preventive maintenance situation 223
pricing-to-win 99
process metrics, quality 64–65
product backlog, scrum process 37
product metrics, quality 64
professional advancement 5–6
professionalism
    abilities 240
    accreditation 239–240
    certification 241
    code of ethics and conduct 243–245
    components *238*
    definition of 238
    legal issues 247–250
    licensing 240–241
    societies of 241–242
    software development standards 245–247
    software engineering programs criteria 240
project management
    definition 91–92
    earned value tracking 103–104
    functional structure type **95**
    key activities **93–94**
    matrix structure type **96**
    organizational structures **95–96**
    planned value (PV) 103–104
    planning activities 96–97
    project structure type **95**
    risk management 105–109
    software estimation 97–101
    task plan and schedule 105
    work-breakdown-structure (WBS) 102, *102*
project metrics, quality 64
project risks 106
project team formation; *see also* team building
    qualification and experience 10–12
    team problems 12, 17–20
proof of concept 3
proxy **187**
pseudocode 209–211
Putman, L. 101

quality assurance *see* software quality assurance
        (SQA)

quality attribute
    definitions 125
    modules 126–127
    purpose of 126
    reliability, DH 127
quality manager 9
quality metrics
    during development 63–65
    post release 65

rational unified process 35
re-engineering process 229
refactoring, extreme programming 37
regression test 76
reliability, quality metrics 65
requirements engineering
    agreement of 124
    analysis phase tasks **124**
    analyzing 132–137
    characteristics 127–128
    conceptual design 132
    context diagram 132
    definitions 125
    eliciting 129–132
    emergent requirement 125–126
    functional *vs.* non-functional 125
    interviews 130
    prototyping 131
    quality attribute 125–128
    role playing 130
    shadowing 130–131
    specifying 137–141
    storyboarding 131
    use case modeling 131, 133–136
    validating 141–143
requirements specification 148–149, 289–298
requirements traceability matrix (RTM) 141, 143
retirement 227
retrospective meeting 273–274
risk management
    analysis 106–108
    exposure 106–107, *107*
    identification 106
    planning and monitoring 109
Royce, W. 31

scrum development process
    backlog items 261
    backlog list 261–264
    burndown chart 270, 271, *271*
    cross-functional team 256
    daily scrum 38, 270

epics 261
estimation process 264–265
implementation phase 259, 266–272
initiation phase 258–259
kaizen 274
key roles 256–257
meeting types 256
ordering method 264–266, *265*
planning meeting 37
planning poker technique 264, *265*
principles 256
product backlog 37
product backlog item (PBI) 261
product owner 256–257
refinement/grooming, backlog 263
reflection phase 259–260, 272–274
relationship of *262*
retrospective 273–274
review 37, 259–260, 272–274
scrum master 256–257
sprint and planning of 37, 266–272
stories 261
team 257
themes 261
tracking progress 270–271
T-shaped person 257
T-shirt sizing technique 264
secure coding 56, 214, **215**
security design 56
security engineering
  key techniques 55–56
  and quality assurance 55
  security attacks 56, 79–80
security requirements engineering 55–56
security risk management 55
security test 77
security verification 56
service-oriented architecture (SOA) 170, *170*
Siemens four-views architecture 165–166
simple object-based architectural
    design(SOBAD) **178–180**
singleton **187**
small releases, extreme programming 36
software configuration management (SCM)
  configuration identification 113–114
  control management 114–115
  elements of 112–113
  importance 110–111
software development life cycle (SDLC) 30–34
software engineering
  advancement 5–6
  changeability 5

complexity 5
component developer 10
conformity 5
defects 4
definition 3–4
development team, role 8–10
foundations 6
invisibility 5
knowledge areas **6–8**
nature 4–5
planning manager 10
quality manager 9
software analyst 8–9
software architect 9
SWEBOK **6–8**
team leader 8
software engineering body of knowledge
    (SWEBOK)
  maintenance of 222
  roles and responsibilities **6–8**
  SCM process 112–113
  software design 153
  software engineering management 91
  software testing 72
  system requirements 125
software engineering code of ethics and
    professional practice (SE Code) 5,
    244–245
software engineering institute (SEI) 5, 34, 38,
    164–165
software engineering process
  agile processes 29–30
  conceptual design 42–43, *44*
  context design 42–43, *43*
  development strategy 42–44, 46–47
  evaluation 39, 45–46
  extreme programming (XP) 35–37
  fundamentals 29
  life cycle models 30–34
  plan-driven processes 29–30
  scrum process 37–38, *38*
  team software process (TSP) 38–39, *39*
  unified process (UP) 35
software estimation
  analogy 99
  business-driven techniques 99
  constructive cost model (COCOMO)
    100–101
  Delphi method 98–99
  estimating to available capacity 99
  expert judgment method 98
  fundamental principles 97

software estimation (*cont.*)
linear regression estimation 99–100, *100*
pricing-to-win 99
project scheduling 102–103, *103*
size estimates 97–98
software life cycle management (SLIM)
100–101
statistical/parametric methods 99–101
software life cycle management (SLIM)
100–101
software quality assurance (SQA)
characteristics, measurement strategy 61
code review 72, 84–85
cost and project schedule 53–54
cost of quality 57–58
cost, requirement defect removal **67**
data analysis 59
defect life cycle 56
defect rate 53
development phase metrics 63–65
goal, question , metrics (GQM) technique
61–63, 82–83
IEEE standard, SQAP 59–60, 81–82
importance 52
inspection 69–72, 83–84
leaving defects 54
managers role 52–53
oversight activity 58
planning activity 58
post release metrics 65
processes associated 58–59
quality measurement and defect tracking
60–61
quality reviews 67–72
questions, measurement strategy 61, 82
record keeping 58
reliability 53
reporting data 59
requirements phase 53–54
security engineering 55–56
software testing 72–79
technical review 68–69
validation and verification 66
and V-Model SDLC 54–55
walk-through 68
software quality assurance plan (SQAP) 59–60,
81–82
software requirements specification (SRS); *see
also* requirements engineering
natural language **140**
organization 138–139
purpose and value 138

writing requirements, guidelines 139
software & systems engineering standards
committee (S2ESC) 245
software testing
development tests types *73*
importance 72
integration testing 75–76, 79, 86–87
successful test implies 73
system testing 76–79
test harness 73
unit testing 74–75, 79, 85–86
use case test planning 79, 88–89
vs. debugging 72
special interest group on software engineering
(SIGSOFT) 242
spiral model *31*, 31–32
sprint, scrum process
backlog 267–268
burnup chart 271
capacity 269
goal 37
planning 267–270, 276–277
retrospective 273–274
review 272–273
scrum board 269
velocity 268–269
stakeholders 162
standards, software development
IEEE-CS standards **246–247**
importance 245
state **188**
state machine diagrams 176–177
static evaluation 66
strategy **188**
stress test 77
structured hierarchical style, SRS **140**
SWEBOK *see* software engineering body of
knowledge (SWEBOK)
system testing 76–79, *77, 78*

task-Gantt chart 102–103, *103*
team building
case study 45
communication skills 27
decision-making procedure 27 28
development activities 28
effective team qualities 25, 45
goal setting 26
ineffective team qualities 25
problems 12, 17–20
roles and responsibilities **8–10,** 26
work control 28

team formation 3
team leader **8**
team problems
    case study 17–20
    description 12
    labor avoidance 19
    quality issue 18
    reporting time 19
team software process (TSP) 38–39, *39*
technical review 68–69
technical risks 106
template method **188**
test-driven development (TDD) 37, 209
trade secret 248
traffic management system (TMS)
    Gantt chart 103
    modularity, design 153, *154*
    risk management 105–108
    state machine diagrams 176–177
    use case modeling *134, 135*
    work breakdown structure 102
transition, unified process 35
Turner, R. 29

unified modeling language (UML) 133
unified process (UP) 35
unit testing techniques
    anomaly analysis 75
    case study 85–86
    categories 74
    definitions of 74
    static techniques 74
    test plan 75
usability test 77
use case, DH system

configuration 305–306
control DH environment 327–329
definition of 131
diagram *303*
humidistat set plan parameters 316–318
login/logout 311–312
manual parameters set 329–330
modify/view, month plan 321–323
monitor sensor values 325–326
monthly planner management 312–315
month report preparation 323–325
operation state change 310
power switch set plan parameters 319–321
read sensor values 326–327
security set plan parameters 318–319
set default parameters 308–309
temperature set plan parameters 315–316
test plan *77–78*, 88–89
user accounts management 307–308
users 302

validation and verification (V&V)
    quality assurance 66
    software requirements activity 141–143
visitor **188**
V-model *32*, 32–33
volume test 77

walk-through, quality review 68
waterfall model *30*, 30–31
white box techniques, unit testing 75
work-breakdown-structure (WBS) 102, *102*
writing requirements
    language requirement 139–141
    styles used 140

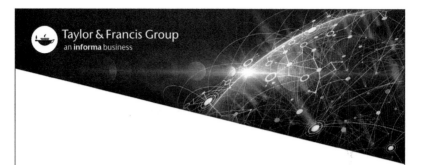